代数

Algebra

中英双语指南 A Bilingual Guide

Table of Contents

前言 Preface

This bilingual book, in English and Chinese, is for people who wish to learn Algebra in both languages. Students from China studying abroad may find that the bilingual instructions make it much easier to understand and master math concepts. On the other hand, people who wish to study or work in China will find the book helpful for learning math vocabulary in Chinese.

这本中英文双语书适合那些希望用两种语言学习代数的人。来自中国的留学生可能会发现双语解释更容易理解和掌握数学概念。另一方面，希望在中国学习或工作的人会发现这本书对学习中文数学词汇很有帮助。

This book will be a supplement to a textbook. We assume readers already have basic language skills in both languages. So the goal is to make it easy and efficient for the reader to learn math vocabulary in another language. We make everything brief to save the reader time. Readers will find it very easy to use this book to quickly review the most important concepts and skills in Algebra.

本书将作为教科书的补充。我们假设读者已经具备两种语言的基本语言技能。因此，我们的目标是让读者轻松高效地学习另一种语言的数学词汇。我们尽量简明扼要，以节省读者的时间。读者会发现使用本书可以非常轻松地快速复习代数中最重要的概念和技能。

This book covers a lot of ground, including mathematics learned in middle school and high school, such as Algebra, Geometry, Statistics, and up to AP Precalculus. It is the second book in a series of Chinese-English bilingual books. The first book, Math Foundations, covers the math learned in elementary school. The series will expand to a wide range of topics, including Calculus, Physics, and other areas. Interested readers, please visit our website for more information.

这本书涵盖了很多内容，包括初中和高中学习的数学，如代数、几何、统计学、直至 AP 初级微积分。这是中英双语书籍系列的第二本。第一本书，《数学的基础》，涵盖了小学学到的数学。该系列将扩展到广泛的主题，包括微积分、物理和其他领域。有兴趣的读者请访问我们的网站了解更多信息。

1   引言 Introduction

代数本质上是用符号代替数字的数学。最简单的如数字的表达式 $5+3$ 可以用符号来表达 $a+b$，这里 $a=5$，$b=3$。

用符号代表未知数使我们可以在一些条件不确定的情况下分析问题，甚至进一步求解未知数。即使是已知的数据，符号也可以代表众多或者复杂数据，让分析变得简单明了。

代数更可以帮助我们将复杂的问题简单化，研究事物的规律，并归纳成简洁的公式。比如爱因斯坦便是用代数推导出他的著名定律

$$E = mc^2,$$

从而发现了能量 E 与物体的质量 m 以及光速 c 之间的密切联系。

Algebra is essentially mathematics that uses symbols instead of numbers. The simplest is a numerical expression $5+3$ can be replaced using symbols $a+b$，here $a=5$，$b=3$.

Representing unknowns with symbols allows us to analyze problems where some conditions are not determined, and even solve for unknowns. Even for known data, symbols can represent multiple or complex data, making analysis simple and straightforward.

Algebra can also help us simplify complex problems, study the laws of objects, and summarize them into concise formulas. For example, Einstein used algebra to derive his famous law

$$E = mc^2,$$

thus discovered the close relationship between energy E, the mass of the object m and the speed of light c.

本书的内容包含中学和高中 Algebra I，Algebra II，和 Precalculus，我们将从易到难一一简单介绍，重点是介绍代数中用到的词汇。今天，代数的应用非常广泛，我们将尽量说明最基本的概念，以帮助读者将来进一步学习更高深的数学。

The content of this book includes middle school and high school Algebra I, Algebra II, and Precalculus. We will briefly introduce them one by one from easy to difficult, focusing on the vocabulary used in algebra. Today, Algebra is widely used in many fields, and we will try to explain the most basic concepts clearly to help readers progress to more advanced mathematics in the future.

2   算术 Pre-algebra

数的分类 Types of Numbers

Table 1: Number types. 数的分类。

Real Numbers 实数 1,2,3 , ... − 5,0,3/8 , 2.3789 , 1.4545 ... , $\sqrt{5}$, π ,	Rational Numbers 有理数 − 5,1/2 , 0,7,0.6 , 2.3333 ...	Integers 整数 ... , − 3 , − 2 , − 1 , 0,1,2,3 , ...	Whole Numbers 非负整数 0,1,2,3 , ...	Natural Numbers[1] 自然数/正整数[1] 1,2,3 , ...
				Zero 零 0
			Negative Numbers 负整数 − 1 , − 2 , − 3 ...	
		Non−integer Fractions 非整数分数 − 3/7 , 125/37 , 11/3		
	Irrational Numbers 无理数 π , $\sqrt{2}$, $5^{1/3}$, e			

[1] 自然数有两种定义方法，一种包括 0，一种不包括 0。 There are two definitions of natural numbers: one includes 0, the other does not.

运算Operations

Table 2: List of operations. 运算列表。

Addition 加	Addend + Addend = Sum 加数 + 加数 = 和	$a + b = c$
Subtraction 减	Minuend – Subtrahend = Difference 被减数 – 减数 = 差	$a - b = c$
Multiplication 乘	Multiplier × Multiplicand = Product 被乘数 × 乘数 = 积	$a * b = c$
Division 除	Dividend ÷ Divisor = Quotient 被除数 ÷ 除数 = 商	$a \div b = c$
	Numerator / Denominator = Fraction 分子 / 分母 = 分数	$\frac{a}{b} = c$
Modulo 模除	Dividend % Divisor = Remainder 被除数 % 除数 = 余数	$a \% b = c$
Exponentiation 乘方	Base ^{Exponent} = Power 底数 ^{指数} = 幂	$a^b = c$

运算的次序 *Order of Operations*

运算的次序遵守以下规则：

（1）括号优先，如有多重括号，里面的括号优先。

（2）乘方、乘和除、加和减的运算顺序为乘方第一，乘和除次之，最后为加和减。

（3）同样的运算，从左到右。

The order of operations follows the following rules:

(1) Parentheses take precedence. If there are multiple parentheses, the innermost parentheses take precedence.

(2) The order of operations of exponentiation, multiplication and division, addition and subtraction is: exponentiation first, multiplication and division second, and addition and subtraction last.

(3) The operations of the same level, proceed from left to right.

运算定律Number Properties

交换律：改变运算的两数位置不改变运算的结果。交换律适用于加法和乘法，但不适用于减法和除法。

Commutative property: Two numbers that are operated on can swap positions without changing the result. The commutative property is applicable to addition and multiplication, but not to subtraction and division.

$$a + b = b + a , a * b = b * a.$$

结合律：如果运算有三个或更多的数，运算的次序无关紧要。结合律同样适用于加法和乘法，但不适用于减法和除法。

Associative Property: When three or more numbers are operated on, the order of operations is irrelevant. The associative property is applicable to addition and multiplication, but not to subtraction and division.

$$a + b + c = (a + b) + c = a + (b + c) \, ;$$
$$a * b * c = (a * b) * c = a * (b * c) \, .$$

分配律：它定义了加法和减法的乘法如何拆分。

Distributive Property: It defines how multiplication is distributed over addition and subtraction

$$a * (b + c) = a * b + a * c \, ;$$
$$a * (b - c) = a * b - a * c \, .$$

恒等性：在二元运算中，如果一项是单位元（恒等元），则运算结果将与另一项相同。加法和减法的单位元都是 0；乘法和除法的单位元是 1。

Identity Property: In a binary operation, if one term is the identity element, the result of the operation will be the same as the other term. The identity element for addition and subtraction is 0; the identity element for multiplication and division is 1.

$$a + 0 = a \, ;$$
$$a - 0 = a \, ;$$
$$a * 1 = a \, ;$$
$$a / 1 = a \, .$$

指数性质 Exponent Properties

$$a^0 = 1 \, ;$$
$$a^m \, a^n = a^{m+n} \, ;$$
$$a^m / a^n = a^{m-n} \, ;$$
$$(a^m)^n = a^{mn} \, ;$$
$$(ab)^m = a^m \, b^m \, .$$

因子分解Factoring

算术基本定理：任何大于 1（$n > 1$）的自然数 n，如果不是素数，都可以唯一地分解为有限个素数的乘积，

$$n = p_1^{n_1} p_2^{n_2} \cdots p_k^{n_k} = \prod_{i=1}^{k} p_i^{n_i},$$

这里 $p_1 < p_2 < ... < p_k$ 是素数，n_i 是正整数，$k > 1$。

例如

$1500 = 2^2 \cdot 3 \cdot 5^3$,

$2040 = 2^3 \cdot 3 \cdot 5 \cdot 17.$

The Fundamental Theorem of Arithmetic: Any natural number n greater than 1 ($n > 1$), if it is not a prime number, can be uniquely decomposed into the product of a finite number of prime numbers,

$$n = p_1^{n_1} p_2^{n_2} \cdots p_k^{n_k} = \prod_{i=1}^{k} p_i^{n_i},$$

where $p_1 < p_2 < ... < p_k$ are primes and the n_i are positive integers, $k > 1$.

For example,

$1500 = 2^2 \cdot 3 \cdot 5^3$,

$2040 = 2^3 \cdot 3 \cdot 5 \cdot 17.$

任何自然数 n ($n > 1$) 要么是素数（只能被 1 和它本身整除），要么是合数（素数的乘积）。

整数因式分解可用于计算最大公约数 (GCD) 和最小公倍数 (LCM)。

Any natural number n ($n > 1$) is either a prime number (can only be divided by 1 and itself) or a composite number (a product of primer numbers).

Integer factoring can be used to calculate the greatest common divisor (GCD), and the least common multiple (LCM).

$$\begin{aligned}
gcd\left(1500,2040\right) &= gcd\left(2^2 \cdot 3 \cdot 5^3, 2^3 \cdot 3 \cdot 5 \cdot 17\right) \\
&= 2^2 \cdot 3 \cdot 5 \\
&= 60, \\
lcm\left(1500,2040\right) &= lcm\left(2^2 \cdot 3 \cdot 5^3, 2^3 \cdot 3 \cdot 5 \cdot 17\right) \\
&= 2^3 \cdot 3 \cdot 5^3 \cdot 17 \\
&= 51000.
\end{aligned}$$

因式分解经常用于简化分数以及计算分数的加法或减法。例如，

Factoring is frequently used to simplify fractions and to calculate fraction addition or subtraction. For example,

$$\frac{1500}{2040} = \frac{2^2 \cdot 3 \cdot 5^3}{2^3 \cdot 3 \cdot 5 \cdot 17} = \frac{5^2}{2 \cdot 17} = \frac{25}{34},$$

$$\begin{aligned}
\frac{1}{1500} - \frac{1}{2040} &= \frac{1}{2^2 \cdot 3 \cdot 5^3} - \frac{1}{2^3 \cdot 3 \cdot 5 \cdot 17} \\
&= \frac{2 \cdot 17}{2^2 \cdot 3 \cdot 5^3 \cdot 2 \cdot 17} - \frac{5^2}{2^3 \cdot 3 \cdot 5 \cdot 17 \cdot 5^2} \\
&= \frac{34 - 25}{2^3 \cdot 3 \cdot 5^3 \cdot 17} = \frac{3}{2^3 \cdot 5^3 \cdot 17} = \frac{3}{17000}.
\end{aligned}$$

表达式，方程和函数 Expression, Equation and Function

表达式是数字、运算符和变量（未知数）的组合，例如

Expression is a combination of numbers, operators, and variables (indeterminate), for example,

$$3.2 + 4.6 \, ,$$
$$2\,x - 5 \, ,$$
$$\pi\,r^{2} \, .$$

方程是包含变量的等式，即两个表达式相等。例如，

An equation is an equality containing variables, i.e., two expressions are equal. For example,

$$2\,x - 5 = 10 \, ,$$
$$\pi\,r^{2} = 15 \, ,$$
$$x^{2} + 2\,x + 1 = 0 .$$

函数是变量之间的一一对应关系。例如，

Functions are one-to-one correspondence between variables. For example,

$$y = 2\,x - 5 \, ,$$
$$y = x^{2} + 2\,x + 1 \, ,$$
$$E = m\,c^{2} .$$

函数经常写成 *f(x)* 的形式，例如，

Functions are frequently written in the form *f(x)*, for example,

$$f(x) = 2\,x - 5 \, ,$$
$$f(a, b) = a + b .$$

3　线性方程 Linear Equations

一元一次方程 Linear Equation in One Variable

一元一次方程有以下形式

$a\,x+b=0\,,$

这里 x 是变量，$a\,(a\neq 0)$ 和 b 是常数。方程的解是

$$x=-\frac{b}{a}\,.$$

A linear equation in one variable can be written in the form

$a\,x+b=0\,,$

where x is a variable, $a\,(a\neq 0)$ and b are constants. The solution of the equation is

$$x=-\frac{b}{a}\,.$$

例子：摄氏度和华氏度的转换
室温为 25 摄氏度，华氏温度是多少？
摄氏温度(°C) =（华氏温度 (°F) − 32) * 5/9。假设华氏温度为 x，则

$$\left(x-32\right)*\frac{5}{9}=25\,,$$

$$x-32=25*\frac{9}{5}=45\,,$$

$$x=45+32=77.$$

答案是 25°C 等于 77 °F。

Example: Celsius and Fahrenheit Conversion
The room temperature is 25 Celsius, what is the temperature in Fahrenheit?

Temperature in degrees Celsius (°C) = (Temperature in degrees Fahrenheit (°F) − 32) * 5/9.

Assume the temperature in Fahrenheit is x, then

$$(x - 32) * \frac{5}{9} = 25,$$

$$x - 32 = 25 * \frac{9}{5} = 45,$$

$$x = 45 + 32 = 77.$$

The answer is 25°C is the same as 77 °F.

绝对值方程Absolute Value Equations

绝对值方程
$|ax + b| = c \, (a \neq 0，c \geqslant 0)$
相当于两个方程
$a x + b = c，a x + b = - c.$
方程有两个解: $x_1 = (c - b)/a，x_2 = (- c - b)/a.$
并非每个方程都有解。在上面的例子中，如果 $c < 0$ 那么方程对于任何 x 不成立，即它没有任何解。

The absolute value equation
$|ax + b| = c \, (a \neq 0，c \geqslant 0)$
is equivalent to two equations
$a x + b = c，a x + b = - c.$
It has two solutions: $x_1 = (c - b)/a，x_2 = (- c - b)/a.$

Not every equation has solutions. In the above example if $c < 0$ then the equation will be false for any x, i.e., it does not have any solution.

在下面的例子中，方程的等号两边都是绝对值:
$|x + 2| = |x - 4|.$
通常绝对值方程有多个解，但这个特殊的例子只有一个解。去除绝对值符号，方程可以重写为两个:
$x + 2 = x - 4，2 = - 4，$ 不成立 ，所以方程无解 。
$x + 2 = - (x - 4)，x + 2 = - x + 4，2 x = 2，x = 1。$

18

因为两个方程中的一个没有解，原本的绝对值方程只有一个解 $x = 1$。

In the following example, both sides of the equation are absolute values:
$$|x + 2| = |x - 4|.$$
Usually an absolute value equation has more than one solution, but for this particular case it has only one solution. To rewrite the equation without absolute value symbols, we have two equations:

$x + 2 = x - 4$, $2 = -4$, *which is false*, *so no solution*.
$x + 2 = -(x - 4)$, $x + 2 = -x + 4$, $2x = 2$, $x = 1$.

Because one of the two equations does not have any solution, the original absolute value equation has only one solution $x = 1$.

二元一次方程 Linear Equation in Two Variables

二元一次方程可以写成
$$a x + b y = c,$$
这里 x y 是变量，a $(a \neq 0)$，b $(b \neq 0)$，和 c 是常数。上面的方程有无穷多的解。

对应每一个 x, 它有一个唯一解 y,
$$y = -\frac{a}{b} x + \frac{c}{b}.$$
y 是 x 的函数。x 叫做自变量，y 叫做因变量。

A linear equation in two variables can be written in the form
$$a x + b y = c,$$
where x y are variables, a $(a \neq 0)$, b $(b \neq 0)$, and c are constants. The above equation has infinite many solutions.

For each x, it has a unique solution y,
$$y = -\frac{a}{b} x + \frac{c}{b}.$$
y is now a function of x. x can now be called an independent variable and y is called a dependent variable.

4   图示 Graphing

一维 One Dimension

任何实数都可以用直线上的一个点表示，这条直线称为轴。例如，下图中，$x = -5$用x轴上位置-5处的点表示；$x = 4$由x轴上位置4处的点表示。

Any real number can be represented as a point on a straight line, called an axis. For example, in the picture below, $x = -5$ is represented by a point at the position -5 on the x-axis; $x = 4$ is represented by a point at the position 4 on the x-axis.

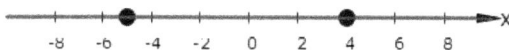

Figure 1: Real numbers represented by dots on the x-axis. x 轴上的点代表实数。

笛卡尔系统 Cartesian System

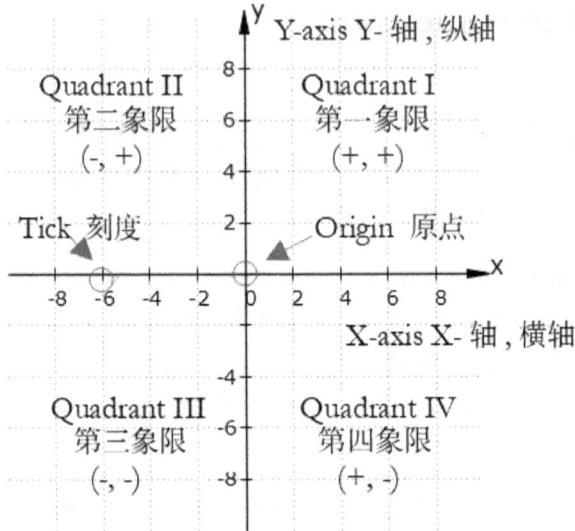

Figure 2: Cartesian system, axis and quadrants. 笛
卡尔系统，坐标轴，和象限。

　　笛卡尔坐标系是用于描述点在空间中的位置的坐标系。在二维中，所有点都在一个平面内。二维笛卡尔坐标系有两个相互垂直的坐标轴：x 轴和 y 轴。两个坐标轴的交点称为原点。这两个轴将平面分为四个区域，逆时针方向，称为第一象限到第四象限。二维平面中的每个点都可以由一对数值确定。这些值对应于 x 轴和 y 轴上的坐标，写成 (x, y)。原点的坐标为 $(0, 0)$。x 轴上的点的坐标为 $(x, 0)$；y 轴上的点的坐标为 $(0, y)$。

Cartesian system is a coordinate system used for specifying a point's position in space. In 2-dimension, all points are within a plane. 2D Cartesian system has two coordinates, x-axis and y-axis, that are perpendicular to each other. The intersection point of the two axes is called the origin. The two axes divide the plane into four areas, counterclockwise, called quadrant I to quadrant

IV. Each point in the 2D plane, can be specified by a pair of values. These values correspond to the coordinates on the *x*-axis and *y*-axis, denoted as (*x, y*). The coordinates for the origin are (0, 0). A point on the *x*-axis has the coordinate (*x*, 0); a point on the *y*-axis has the coordinate (0, *y*).

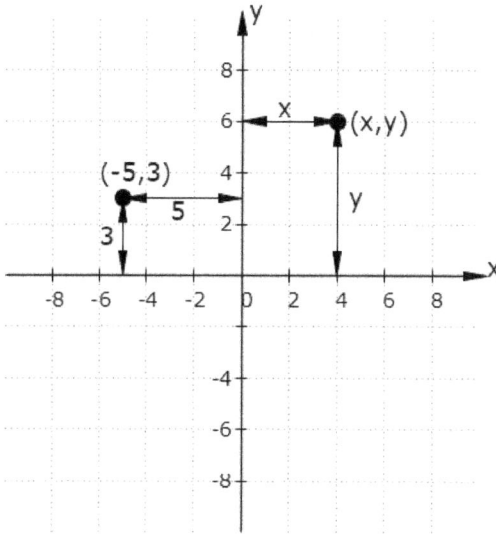

Figure 3: Points in a 2D plane represented by a pair of real numbers in the Cartesian system. 二维平面中的点由笛卡尔系中的一对实数表示。

线性方程绘图 Graphing Linear Equation

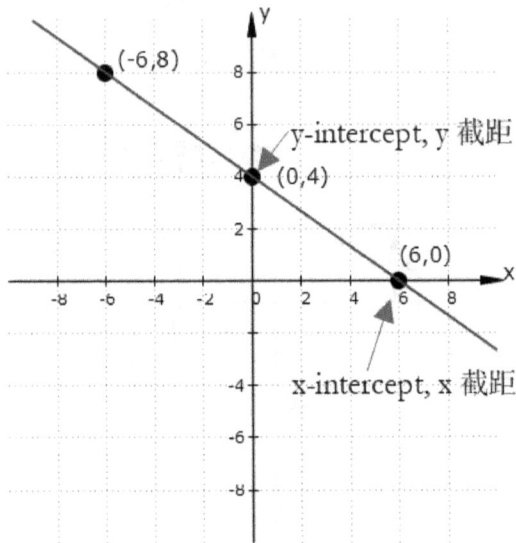

Figure 4: A linear equation is represented by a straight line in the Cartesian system. 线性方程在笛卡尔系统中是一条直线。

线性方程 $ax + by + c = 0$ 的一个解，x 和 y, 在笛卡尔坐标系中可以用一个点 (x, y) 代表。方程的所有解构成一条直线。在上图中（图 4），我们画出了线性方程 $2x + 3y = 12$，大家可以验证直线上的三个点 $(6, 0)$, $(0, 4)$, 和 $(-6, 8)$ 都是方程的解。

在 $a = 0$ $(b \neq 0)$ 的特殊情形，线性方程变成 $y = -\dfrac{c}{b}$。图形是一条水平直线，y-截距为 $-\dfrac{c}{b}$。如果 $b = 0$ $(a \neq 0)$, 方程为 $x = -\dfrac{c}{a}$，图形是一条垂直直线，x-截距为 $-\dfrac{c}{a}$。

A solution to the linear equation $a\,x+b\,y+c=0$, x and y, can be represented by a point (x, y) in the Cartesian system. All solutions then form a straight line. In the graph above (Figure 4), we draw the linear equation $2\,x+3\,y=12$, one can verify the three points on the line $(6, 0)$, $(0, 4)$, and $(-6, 8)$ are all solutions to the equation.

In the spacial case of $a=0$ $(b\neq 0)$, the linear equation becomes $y=-\dfrac{c}{b}$. The graph is a horizontal line with y-intercept at $-\dfrac{c}{b}$. On the other hand, if $b=0$ $(a\neq 0)$, the equation is $x=-\dfrac{c}{a}$, the graph is a vertical line with x-intercept at $-\dfrac{c}{a}$.

斜截式 Slope-intercept Form

我们可以将方程 $a\,x+b\,y=c$ 重写成以下形式

$$y=-\frac{a}{b}\,x+\frac{c}{b}\ 。$$

更简单的形式是

$$y=m\,x+y_0\ ，$$

这里 $m=-\dfrac{a}{b}$ 是斜率，$y_0=\dfrac{c}{b}$ 是 y 截距。方程也可以用斜率和 x 截距写成下列形式

$$y=m\left(x-x_0\right),$$

这里 $x_0=\dfrac{c}{a}$ 是 x 截距。

We can rewrite the equation $a\,x+b\,y=c$ in the form of

$$y = -\frac{a}{b}x + \frac{c}{b}.$$

In a simpler form, it is

$$y = mx + y_0,$$

where $m = -\frac{a}{b}$ is the slope, $y_0 = \frac{c}{b}$ is the y-intercept. It can also be rewritten using slope and x-intercept as

$$y = m\left(x - x_0\right),$$

where $x_0 = \frac{c}{a}$ is the x-intercept.

截距式 Intercept Form

如果一直线有非零的 x 截距 x_0 和非零的 y 截距 y_0 , 线性方程是

If a line has both non-zero x-intercept x_0 and non-zero y-intercept y_0, the linear equation is

$$\frac{x}{x_0} + \frac{y}{y_0} = 1.$$

点斜式 Point-slope Form

非垂直的直线斜率不是零 $m \neq 0$。这样的直线可以由斜率 m 和线上的一点 (x_1, y_1) 决定。直线是

A non-vertical line has a non-zero slope $m \neq 0$. The line can be defined by the slope m and a point (x_1, y_1) on the line. The line is

$$y = y_1 + m\left(x - x_1\right),$$
$$y - y_1 = m\left(x - x_1\right).$$

两点式 Two-point Form

一个连接两点 (x_1, y_1) 和 (x_2, y_2) 的直线，如果 $x_1 \neq x_2$，方程是

For a line that passes through two points (x_1, y_1) and (x_2, y_2), if $x_1 \neq x_2$, then the equation is

$$y - y_1 = m\left(x - x_1\right),$$
$$m = \frac{\left(y_2 - y_1\right)}{\left(x_2 - x_1\right)}.$$

5 不等式Inequalities

单变量不等式 One Variable Inequality

不等式是两个表达式之间的比较。该陈述判断（命题）要么是真，要么是假。如果不等式有未知变量，则求解不等式就是找到使原始命题成立的解。

An inequality is a comparison between two expressions. The statement is either true or false. If an inequality has unknown variables, solving the inequality is finding the solutions that make the original statement true.

Table 3: List of inequalities. 不等式的种类。

Inequality Symbols	English	中文
<	less than	小于
>	greater than	大于
≤ or <=	not greater than, less than or equal to	不大于，小于等于
≥ or >=	not less than, greater than or equal to	不小于，大于等于

解不等式的过程与解方程类似。以下是一个示例：

The procedure for solving an inequality is similar to solving an equation. Showing below is an example：

$$2x + 3 < 9,$$
$$2x < 9 - 3,$$
$$2x < 6,$$
$$x < 3.$$

不等式的解 $x < 3$ 可以图示，在 x 轴上将小于 3 的范围画上阴影（图 5）。请注意，在 $x = 3$ 处，我们用了一个空心圆，这是因为端点不是解的一部分 ($x < 3$)。如果它是解的一部分（$x \leq 3$），我们就会使用实心圆。

The solution $x < 3$ can be graphed on x-axis by shading the range that is smaller than 3 (Figure 5). Note at $x = 3$ we have an open circle, that is because the endpoint is not part of the solution ($x < 3$). Had it been part of the solution ($x \leq 3$), we would have used a closed circle instead.

Figure 5: Inequality (x < 3) can be represented by a shaded arrow and an open circle. 不等式 (x < 3) 可以用带阴影的箭头和空心圆来表示。

绝对值不等式Absolute Value Inequalities

举个例子，我们来解不等式

$|x - 2| \leq 3$ 。

绝对值不等式等价于两个不等式，

$x - 2 \leq 3$ 和 $x - 2 \geq -3$ ，

可以写成一行

$-3 \leq x - 2 \leq 3$ 。

其解为

$-3 + 2 \leq x \leq 3 + 2$,
$-1 \leq x \leq 5$ 。

As an example, we try to solve the inequality

$|x - 2| \leq 3.$

It is equivalent to two inequalities,

$x - 2 \leq 3 \text{ and } x - 2 \geq -3$,

can be written in one line as

$-3 \leq x - 2 \leq 3.$

The solution is then

$-3 + 2 \leq x \leq 3 + 2$,
$-1 \leq x \leq 5.$

Figure 6: Inequality -1 ≤ x ≤ 5 can be represented by a shaded segment and closed circles. 不等式 -1 ≤ x ≤ 5 可以用带阴影的线段和实心圆来表示。

不等式

$|x - 2| > 3$,

其解为

$x - 2 < -3$ 或 $x - 2 > 3$,

$x < -1$ 或 $x > 5$ 。

For the inequality

$|x - 2| > 3$,

the solutions are

$x - 2 < -3$ or $x - 2 > 3$,

$x < -1$ or $x > 5$.

Figure 7: Inequalities x < -1 or x > 5 can be represented by two shaded arrows and open circles. 不等式 x < -1 或 x > 5 可以用带阴影的箭头和空心圆来表示。

双变量不等式图示 Graphing Two Variable Inequality

我们用下面的例子来演示怎样图示两个变量的不等式：

$y - 3x < 6$ 。

首先我们在 xy 平面上画一条直线

$y - 3x = 6$ 。

这条直线会将平面分成两半：没有阴影的左边和有阴影的右半边。我们可以验证阴影部分的所有的点都是不等式的解。拿原点（0，0）为例 $y - 3x = 0 - 3*0 = 0$，小于 6。所以有阴影的区域可以用来表示最初的不等式的解。

To demonstrate how to graph a two variable inequality, we use the following example,

$y - 3x < 6$.

First we draw

$y - 3x = 6$

on the *xy*-plane, which is a straight line separating the plane into two parts: unshaded left side and shaded right side. One can verify that all points in the shaded area are solutions to the inequality. Take the origin (0, 0) as an example, *y − 3x = 0 − 3*0 = 0*, which is less than 6. So the shaded area represents solutions to the original inequality.

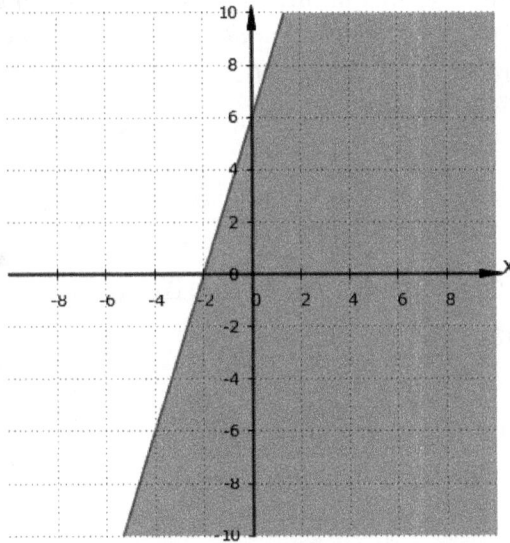

Figure 8: Two variable inequality y − 3 x < 6 can be represented by shaded area in the Cartesian coordinate system. The line separating the two areas is y − 3 x = 6. 二元不等式 *y − 3 x < 6* 可以用笛卡尔系统里阴影部分表示。分割两个部分的直线是*y − 3 x = 6*。

6 线性系统Linear System

线性方程组是一组具有相同变量的线性方程。下面是一个包含两个变量的示例。

A system of linear equations is a set of linear equations with the same variables. Below is an example with two variables.

$$2x + 3y = 12 \qquad (1)$$
$$3x - 2y = 5 \qquad (2)$$

替代法 Solving by Substitution

式（1）可以重写成如下方程

We can rewrite Eq. (1) in the following

$$y = (12 - 2x)/3 \qquad (3)$$

将 y 代入式（2）得到

and substitute y in terms of x back into Eq. (2), we get

$$3x - \frac{2(12 - 2x)}{3} = 5$$
$$3x - 8 + \frac{4}{3}x = 5$$
$$\frac{13}{3}x = 5 + 8$$
$$13x = 13 * 3$$
$$x = 3 \qquad (4)$$

将 x 的解（式（4））代入式（3），我们得到 $y = 2$。方程组的解是 $x = 3$，$y = 2$。

Substitute solution for x (Eq. (4)) back to Eq. (3), we get $y = 2$. So the solution of the linear system is $x = 3, y = 2$.

消元法 Solving by Elimination

方程（1）和（2）可以重写成如下：

Rewrite the original equations (1) and (2) as follows:

$$2(2x + 3y) = 2 * 12 \tag{5}$$
$$3(3x - 2y) = 3 * 5 \tag{6}$$

简化后得到

After simplification, we have

$$4x + 6y = 24 \tag{7}$$
$$9x - 6y = 15 \tag{8}$$

将式（7）和式（8）相加，将消去变量 y，其结果是

Adding Eq.(7) and Eq.(8) together, we eliminate the variable y, the resulting equation is

$$13x = 39.$$

我们找到同样的解 $x = 3$。将 x 值回代式（1），我们得到 $y = 2$。

Again we find the solution $x = 3$. Substituting this value back to the original Eq. (1), we'll find $y = 2$.

绘图法 Solving by Graphing

式（1）和式（2）可以重写成两个函数：

Eq. (1) and Eq.(2) can be rewritten as two functions:

$$y = (12 - 2x)/3 \qquad\qquad (9)$$
$$y = -(5 - 3x)/2 \qquad\qquad (10)$$

在 xy 平面上，它们的图形是两条直线。直线的交点$(3, 2)$将是两个方程的共同解。

Graphing on a xy-plane, they will be two straight lines. The cross point $(3, 2)$ will be the solution to both equations.

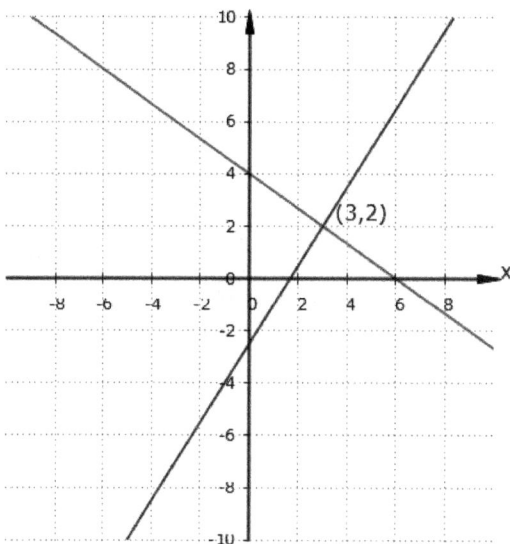

Figure 9: *Solutions to two linear equations is the cross point of two straight lines.* 两个线性方程的解是两条直线的交点。

7 多项式Polynomials

单项式Monomials

Table 4: Examples of monomials. 单项式的例子。

Example 例子	Monomial 单项式	Degree 次数	Rational 原因
-5.5	Yes 是	0	$-5.5x^0$
$2^{1/2}x$	Yes 是	1	Coefficient can be real number 系数可以是实数
$2+t$	No 否		Can only be 1 term 只能一项
$3.2x^5yz^2$	Yes 是	8	Can have multiple variables 可以有多个变量
$1/x$	No 否		$1/x=x^{-1}$, exponent can't be negative 指数不能是负数
$x^{1/2}$	No 否		Exponent must be natural number 指数必须是自然数
3^y	No 否		Variable can't be exponent 变量不能是指数

　　单项式是由数字或者变量的积组成的表达式,表4中列有单项式的例子。变量的指数必须是自然数（包括0）。单独一个数字也是单项式。变量的指数也叫次，如果有多个变量，次是各个变量的指数的和。与变量相乘的常数又叫系数。

A monomial is an expression consisting of the product of a number and variables. Examples are shown in Table 4. The exponent of the variable must be a natural number (including 0).

A single number is also a monomial. The exponent of a variable is also called the degree. If there are multiple variables, the degree is the sum of the exponents of each variable. The constant multiplied by variables is also called the coefficient.

二项式和三项式Binomials and Trinomials

二项式是两个单项式的和，三项式是三个单项式的和。也就是说，二项式有两项，三项式有三项。

Binomials are the sum of two monomials, and trinomials are the sum of three monomials. In other words, binomials have two terms, and trinomials have three terms.

二项式例子 Binomial Examples

$a + b\,x$

$x^2 + y^2$

$ax^m - bx^n$

$0.5\,x^4 + \pi\,t$

三项式例子 Trinomial Examples

$x^2 + y^2 + z^2$

$a\,x^2 + b\,x + c$

$xy + yz + zx$

多项式 Polynomials

多项式有多个单项式的项。单变量 x 的多项式可以写为

$$\sum_{k=0}^{n} a_k\, x^k = a_n\, x^n + a_{n-1}\, x^{n-1} + \cdots + a_2\, x^2 + a_1\, x + a_0 ,$$

这里 $a_k\left(k = 0 \cdots n , a_n \neq 0\right)$是常数，又叫系数。多项式的次数是最大的指数，$n$。

Polynomials have more than one terms of monomials. Polynomials with one variable x can be written as

$$\sum_{k=0}^{n} a_k\, x^k = a_n\, x^n + a_{n-1}\, x^{n-1} + \cdots + a_2\, x^2 + a_1\, x + a_0 ,$$

where $a_k\left(k = 0 \cdots n , a_n \neq 0\right)$ are constants, also called coefficients. The degree of the polynomial is the largest exponent, n.

加法和减法 Addition and subtraction

运用结合律，相同的项可以合并，方法是将项的系数相加减。

Like terms are combined together by adding or subtracting their coefficients using the associative peroperty.

$$
\begin{aligned}
& 2\,x^2 + 3\,x + 1 \\
+\ & \left(3\,x^2 + 5\,x + c\right) \\
=\ & 5\,x^2 + 8\,x + 1 + c
\end{aligned}
$$

乘法 Multiplication

重复运用分配律，可以得到两个多项式相乘的结果。下面是两个例子：

Two polynomials can be multiplied using the distributive law repeatedly. Below are two examples:

$$
\begin{aligned}
x\left(x^2 + 3\,x + 2\right) &= x^3 + 3\,x^2 + 2\,x , \\
\left(x+2\right)\left(x+3\right) &= x\left(x+3\right) + 2\left(x+3\right) \\
&= x^2 + 3\,x + 2\,x + 6 \\
&= x^2 + 5\,x + 6 .
\end{aligned}
$$

除法 Division

两个多项式相除的结果通常不再是多项式。在第八章和第十五章我们再讨论。

Division of two polynomials usually results in an expression that is no longer a polynomial. We'll discus more in Chapter 8 and 15.

8 因子分解Factoring

特例 Special examples

$$a^2 - b^2 = (a+b)(a-b)$$
$$a^2 + 2ab + b^2 = (a+b)^2$$
$$a^2 - 2ab + b^2 = (a-b)^2$$
$$a^2 + b^2 + c^2 + 2ab + 2bc + 2ca = (a+b+c)^2$$
$$a^3 + b^3 = (a+b)(a^2 - ab + b^2)$$
$$a^3 - b^3 = (a-b)(a^2 + ab + b^2)$$

GCF 因子分解 Factoring Using GCF

例子：因子分解 $18 x^2 y + 24 xy^2$。

两项的最大公因子是

$$GCF\left(18 x^2 y, \ 24 xy^2\right) = GCF\left(2 \cdot 3^2 x^2 y, \ 2^3 \cdot 3 x y^2\right)$$
$$= 2 \cdot 3 xy = 6 xy \text{。}$$

最初的表达式可以是 GCF 和剩余因子的乘积

$$18 x^2 y + 24 xy^2 = 6 xy\left(3x + 4y\right) \text{。}$$

Example: Factoring $18 x^2 y + 24 xy^2$.

The greatest common factor of the two terms is

$$GCF\left(18 x^2 y, \ 24 xy^2\right) = GCF\left(2 \cdot 3^2 x^2 y, \ 2^3 \cdot 3 x y^2\right)$$
$$= 2 \cdot 3 xy = 6 xy \text{.}$$

The original expression can be a product of the GCF and remaining factor:

$$18\,x^2\,y + 24\,xy^2 = 6\,xy \cdot 3\,x + 6\,xy \cdot 4\,y = 6\,xy\left(3\,x + 4\,y\right).$$

因子式多项式 Polynomial Equations in Factored Form

多项式方程是等号一侧为多项式而另一侧为 0 的方程。下面是因子形式的多项式方程的一个例子：

$$6\,xy\left(x - 2\right)\left(x - 3\right) = 0 \ 。$$

每个含有变量的因子等于 0，得出方程的解。方程的解为

$x = 0$；
$y = 0$；
$x - 2 = 0$，$x = 2$；
$x - 3 = 0$，$x = 3$ 。

Polynomial equations are equations with polynomials on one side of the equal sign and 0 on the other. Here is an example of a polynomial equation in factor form:

$$6\,xy\left(x - 2\right)\left(x - 3\right) = 0.$$

The solutions are that each factor with a variable equals to 0. So the solutions are

$x = 0$；
$y = 0$；
$x - 2 = 0$，$x = 2$；
$x - 3 = 0$，$x = 3$.

多项式函数图示 Graphing Polynomial Functions

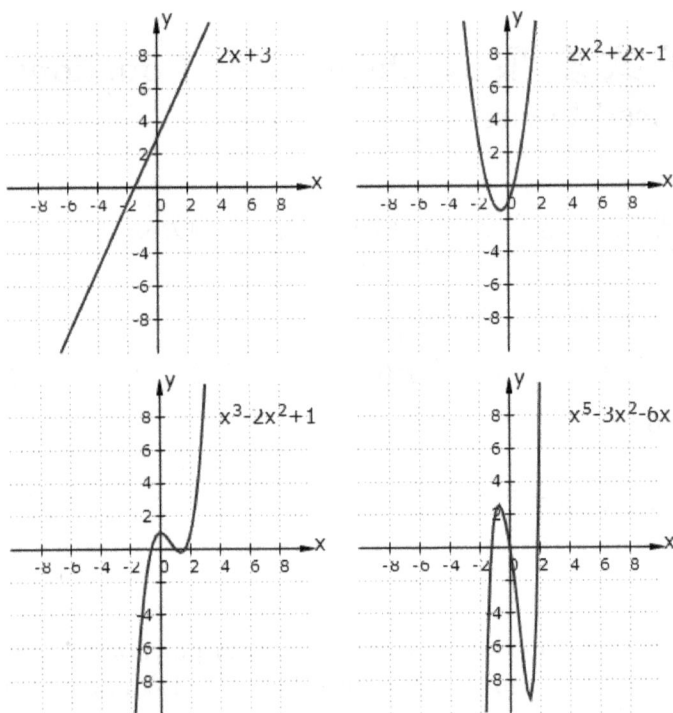

Figure 10: Graphs of polynomial functions of different order. 各种次数的多项式函数图形。

多项式也可以写成函数形式，例如

$f(x) = 2x + 3$,
$g(x) = 2x^2 + 2x - 1$.

一元函数可以在 xy 坐标系里用图形表示（图 10）。

Polynomials can also be written in the function form, for example

$$f(x) = 2x + 3,$$
$$g(x) = 2x^2 + 2x - 1.$$

One variable polynomial function can be graphed on xy-coordinate plane (Figure 10).

9   二次方程Quadratic Equations

一元二次方程的标准形式是

$$a\,x^2 + b\,x + c = 0\,,$$

这里 a, b, c 是常量，且 $a \neq 0$。a, b, c 又叫系数，a 被称为二次系数，b 线性系数，c 常系数。

Quadratic equations are single variable equations of the standard form

$$a\,x^2 + b\,x + c = 0\,,$$

where a, b, c are constants and $a \neq 0$. a, b, c are called coefficients, and separately a is the quadratic coefficient, b the linear coefficient, c the constant coefficient.

因子分解法 Solving by Factoring

因式分解是求解二次方程的一种方法。例如，通过检验以下方程，我们发现 $x = 1$ 是一个解。我们可以进一步分解方程来找到另一个解。

One way to solve a quadratic equation is by factoring. For example by inspecting the following equation, we find $x = 1$ is a solution. We can further factoring it to find the other solution.

$$
\begin{aligned}
2\,x^2 + 3\,x - 5 &= 0 \\
2\,x^2 - 2\,x + 5\,x - 5 &= 0 \\
2\,x\,(x - 1) + 5\,(x - 1) &= 0 \\
(2\,x + 5)\,(x - 1) &= 0
\end{aligned}
$$

$$
\begin{array}{ll}
2\,x + 5 = 0 & \quad x - 1 = 0 \\
x = -2.5 & \quad x = 1
\end{array}
$$

配方法 Complete the Square

配方法首先将二次项和线性项变成线性表达式的平方。下面的例子说明了这一点。

Completing the square method starts with make the quadratic and linear term into a square of linear expression. It is illustrated by the following example.

$$2\,x^2 + 3\,x - 5 = 0$$

$$x^2 + \frac{3}{2}\,x - \frac{5}{2} = 0$$

$$x^2 + 2\cdot\frac{3}{4}\,x + \left(\frac{3}{4}\right)^2 - \left(\frac{3}{4}\right)^2 - \frac{5}{2} = 0$$

$$\left(x + \frac{3}{4}\right)^2 = \frac{49}{16} = \left(\frac{7}{4}\right)^2$$

$$x + \frac{3}{4} = \frac{7}{4} \qquad x + \frac{3}{4} = -\frac{7}{4}$$

$$x = 1 \qquad x = -\frac{5}{2}$$

公式 Formula

配方法也可以用来推导解二次方程的公式。

Completing the square can be used to derive a general formula for quadratic equations.

$$a x^2 + b x + c = 0$$

$$x^2 + \frac{b}{a} x + \frac{c}{a} = 0$$

$$x^2 + \frac{b}{a} x + \left(\frac{b}{2a}\right)^2 - \left(\frac{b}{2a}\right)^2 + \frac{c}{a} = 0$$

$$\left(x + \frac{b}{2a}\right)^2 = \left(\frac{b}{2a}\right)^2 + \frac{c}{a} = \frac{b^2 - 4ac}{4a^2}$$

$$x + \frac{b}{2a} = \pm \frac{\sqrt{b^2 - 4ac}}{2a}$$

$$x = \frac{-b \pm \sqrt{b^2 - 4ac}}{2a}$$

因子分解二次方程的公式是

The formula for factorization of quadratic equation is

$$a x^2 + b x + c = a (x - p)(x - q),$$

$$p = \frac{-b + \sqrt{b^2 - 4ac}}{2a},$$

$$q = \frac{-b - \sqrt{b^2 - 4ac}}{2a}.$$

如果 $\Delta = b^2 - 4ac < 0$，那么公式中的平方根不是实数，也就是说方程没有实数解。$\Delta = b^2 - 4ac$ 称作判别式。

If $\Delta = b^2 - 4ac < 0$, then the square root in the formula is not a real number, that is, the equation does not have real number solutions. $\Delta = b^2 - 4ac$ is called the discriminant.

$$\Delta = b^2 - 4ac \begin{cases} > 0, & 2 \text{ real number solutions 两个实数解} \\ = 0, & 1 \text{ real number solution 一个实数解} \\ < 0, & 0 \text{ number solution 无实数解} \end{cases}$$

最大值最小值 Maximum and Minimum of Quadratic Function

运用配方法，二次方程可以写成下列形式:

Using the same strategy in completing the square, the quadratic equation can be written as following:

$$y = ax^2 + bx + c$$
$$= a\left(x + \frac{b}{2a}\right)^2 + \frac{4ac - b^2}{4a^2}.$$

如果$x = -\frac{b}{2a}$，第一项是0。除了这个点之外，第一项要么总是正值($a > 0$)要么总是负值($a < 0$)。所以$x = -\frac{b}{2a}$时，y有最小值或者最大值。

If $x = -\frac{b}{2a}$, the first term is 0. Away from this point, the first term is either always positive ($a > 0$) or always negative ($a < 0$). As a result y has either minimum or maximum at $x = -\frac{b}{2a}$.

$$If\ a > 0,\ y_{min} = \frac{4ac - b^2}{4a^2},$$
$$If\ a < 0,\ y_{max} = \frac{4ac - b^2}{4a^2}.$$

二次函数图示 Graphing Quadratic Function

二次函数的图形是抛物线形状，它向上 ($a > 0$) 或向下 ($a < 0$) 展开（图11）。y轴截距位于$y = c$。如果 $\Delta > 0$，则有两个 x 截距，它们是二次方程的实数解或实数根。如果 $\Delta = 0$，则图形与 x 轴相切。接触点为两个相同的根。如果 $\Delta < 0$，则图形不与

x 轴相交，方程没有实数解。曲线的最小值或最大值位于
$x = -\dfrac{b}{2a}$ 。

Graph of quadratic function is of parabola shape. It opens either upward ($a > 0$) or downward ($a < 0$) (Figure 11). The y-intercept is at $y = c$. If $\Delta > 0$, there are two x-intercepts, which are real solutions or roots of the quadratic equation. If $\Delta = 0$, the graph is tangent to the x-axis. The point of contact is a double root. If $\Delta < 0$, the graph does not intercept x-axis, the equation does not have real roots. The minimum or maximum of the graph is at $x = -\dfrac{b}{2a}$.

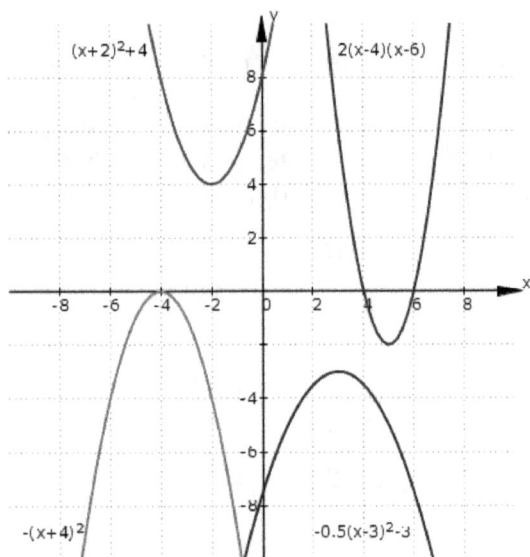

Figure 11: Graphs of quadratic functions.
二次函数的图形。

重力下的运动 Motion Under Gravity

假设一个物体从高处 h 自由落下，需要多长时间才会落到地面？

描述物体垂直下落的公式是一个二次方程：

$$y(t) = v_0 t + y_0 - \frac{1}{2} g t^2$$

这里 $y(t)$ 是物体高度，v_0 是初始速度 (m/s)，y_0 是初始高度 (m)，t 是过去的时间 (s)，g 是重力加速度 $(9.81\ m/s^2)$。

对于我们的题目，$v_0 = 0$，$y_0 = h$，$y(t) = 0$. 方程变为：

$$0 = h - \frac{1}{2} g t^2,$$

$$t^2 = \frac{2h}{g},\ t = \sqrt{\frac{2h}{g}}.$$

Suppose an object is free falling from a height h, how long does it take to land on the ground?

The formula describing the vertical free fall motion is a quadratic equation:

$$y(t) = v_0 t + y_0 - \frac{1}{2} g t^2$$

where $y(t)$ is the height of the object, v_0 is the initial velocity (m/s), y_0 is the initial height (m), t is time elapsed (s), and g is the acceleration due to gravity $(9.81\ m/s^2)$.

For our problem, $v_0 = 0$, $y_0 = h$, $y(t) = 0$. The equation becomes:

$$0 = h - \frac{1}{2} g t^2,$$

$$t^2 = \frac{2h}{g},\ t = \sqrt{\frac{2h}{g}}.$$

10 根号Radicals

函数$f(x)=x^n$称为幂函数，其中n是正整数。当$n \geq 2$，根函数是幂函数的逆函数，它会消去幂。写为：

$$g(x)=x^{\frac{1}{n}} \ \text{ or } \ g(x)=\sqrt[n]{x}.$$

幂函数和根函数互为逆函数，所以：

$$f(g(x))=\left(x^{\frac{1}{n}}\right)^n=x^{\frac{1}{n} \cdot n}=x$$
$$g(f(x))=\left(x^n\right)^{\frac{1}{n}}=x^{n \cdot \frac{1}{n}}=x$$

A function $f(x)=x^n$ is called a power function, where n is a positive integer. For $n \geq 2$, a radical function is the reverse of the power function, it undoes the power. Written as:

$$g(x)=x^{\frac{1}{n}} \ \text{ or } \ g(x)=\sqrt[n]{x}.$$

Power function $f(x)$ and radical function $g(x)$ are inverse function of each other, so:

$$f(g(x))=\left(x^{\frac{1}{n}}\right)^n=x^{\frac{1}{n} \cdot n}=x \ \text{ and }$$
$$g(f(x))=\left(x^n\right)^{\frac{1}{n}}=x^{n \cdot \frac{1}{n}}=x$$

$\sqrt{\ }$称作根号，$\sqrt[n]{\ }$称作n次根，根号下的量称为被开方数。最简单的根式$n=2$，我们简单地使用$\sqrt{\ }$符号，称它为平方根或开平方。

The $\sqrt{\ }$ symbol is called radical symbol or root symbol. $\sqrt[n]{\ }$ is called n-th radical or n-th root, the expression under the symbol is called radicand. For the simplest radical $n=2$, we simply use $\sqrt{\ }$ symbol and call it square root or square radical.

如果 $x\,y$ 全是正实数，

If x y are both positive real numbers,

$$\sqrt[n]{x^n} = x\,,$$
$$\sqrt[n]{x\,y} = \sqrt[n]{x}\;\sqrt[n]{y}\,,$$
$$\sqrt[n]{\frac{x}{y}} = \frac{\sqrt[n]{x}}{\sqrt[n]{y}}\,.$$

如果 x 可能是负数，

If x may be negative,

$$\sqrt[n]{x^n} = \begin{cases} |x| & \text{if } n \text{ is even}\,, \\ x & \text{if } n \text{ is odd}\,. \end{cases}$$

初学者需小心避免一个常见错误，那就是必须首先计算根号下的加减法，然后才能开方.

$$\sqrt[n]{x + y} \neq \sqrt[n]{x} + \sqrt[n]{y}\,.$$

例 如 $\sqrt{5}$ 是 一 个 无 理 数 ， 它 不 能 写 成 $\sqrt{4+1} = \sqrt{4} + \sqrt{1} = 2 + 1 = 3$ 。

A common mistake to avoid for beginners is that one must first calculate additions and subtractions under the radical symbol, only then proceed to the radial operation.

$$\sqrt[n]{x + y} \neq \sqrt[n]{x} + \sqrt[n]{y}\,.$$

For example $\sqrt{5}$ is an irrational number, it can't be written as $\sqrt{4+1} = \sqrt{4} + \sqrt{1} = 2 + 1 = 3.$

平方根函数 Square Root Functions

严格地说平方根函数应该是 $\sqrt{x}$ 。广义地说平方根函数是根号下的表达式含有变量 x，如 $3\sqrt{x+2}+1$ 。

如果变量 x 是正整数，那么 $\sqrt{x}$ 要么是整数，要么是无理数。例如 $\sqrt{2}$ 和 $\sqrt{3}$ 是无理数，但 $\sqrt{4}$ 等于整数 2。

$\sqrt{2}$ 是无理数可以用反证法证明。证明的方法是先假设原命题是错误的，然后推理出明显矛盾，所以原命题是错误的假设不成立，从而得出原命题一定正确的结论。

命题：$\sqrt{2}$ 是无理数。

反证法：

假设 $\sqrt{2}$ 是一个有理数，那么 $\sqrt{2} = \dfrac{p}{q}$，这里 p 和 q 是正整数，最大公约数 $gcd(p,\ q) = 1$。

$$\sqrt{2}^2 = 2 = \left(\dfrac{p}{q}\right)^2, \quad 2\,q^2 = p^2 \text{。}$$

p^2 是偶数，p 必然是偶数，

$$p = 2\,p_1, \quad p^2 = 4\,p_1^2 \text{。}$$
$$2\,q^2 = 4\,p_1^2, \quad q^2 = 2\,p_1^2,$$

因此 $q2$ 也是偶数，所以 q 是偶数。因为 p 和 q 都是偶数，$gcd(p,\ q) \geqslant 2$，这和 $gcd(p,\ q) = 1$ 自相矛盾。所以 $\sqrt{2}$ 是一个有理数的假设不成立，$\sqrt{2}$ 不是一个有理数，也就是说 $\sqrt{2}$ 是一个无理数。

Strictly speaking, the square root function is a function that can be written as $\sqrt{x}$. Loosely speaking, the square root function is a function that involves a square root over an expression containing the variable x, for example, $3\sqrt{x+2}+1$.

If variable x is a positive integer, $\sqrt{x}$ will either be an integer or an irrational number. For example $\sqrt{2}$ and $\sqrt{3}$ are irrational number, but $\sqrt{4}$ equals to integer 2.

That $\sqrt{2}$ is an irrational number can be proving by contradiction. In the proof, one first assume the statement is false, then prove it will lead to a contradiction, therefore the original statement can't be false, so must be true.

Statement: $\sqrt{2}$ is an irrational number.

Proof by Contradiction:

Assume $\sqrt{2}$ is a rational number, then $\sqrt{2} = \dfrac{p}{q}$, where p and q are positive integers, greatest common divisor $gcd(p, q) = 1$.

$$\sqrt{2}^2 = 2 = \left(\frac{p}{q}\right)^2 , \ 2q^2 = p^2.$$

p^2 is even, p must be even,

$$p = 2 p_1 , \ p^2 = 4 p_1^2.$$
$$2 q^2 = 4 p_1^2 , \ q^2 = 2 p_1^2 ,$$

Therefore, q2 is also even, and hence q must be even. Because both p and q are even, $gcd(p, q) \geq 2$, which is contradictory to $gcd(p, q) = 1$. Therefore the assumption is false, we conclude $\sqrt{2}$ is not a rational number, i.e., it must be an irrational number.

平方根函数$\sqrt{x}$的图形是抛物线的一半。另一半不显示是因为我们定义时要求$\sqrt{x}$是正值。

The graph of the square root function $\sqrt{x}$ is a half parabola. The other half is missing because, by definition, we restrict $\sqrt{x}$ to be positive.

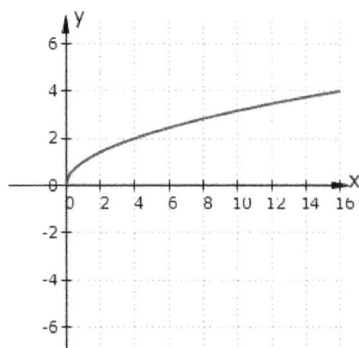

Figure 12: Graph of square root function. 平方根函数图形。

52

立方根函数Cubic Root Functions

立方根$\sqrt[3]{x}$是方程$y^3 = x$的解。如$\sqrt[3]{27} = 3$，因为$3^3 = 27$。

Cubic root $\sqrt[3]{x}$ is a solution of the equation $y^3 = x$. For example $\sqrt[3]{27} = 3$, because $3^3 = 27$.

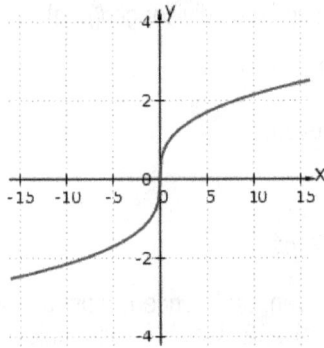

Figure 13: Graph of cubic root function. 立方根函数图形。

根式方程 Radical Equations

根式方程是含有根号，且根号下有变量的方程。下列几个例子：

A radical equation has a variable in the radicand of a radical expression. Some examples are listed below:

$$\sqrt{x+1} - 5 = 0 \, ;$$
$$\sqrt{x+4} - x + 2 = 0 \, ;$$
$$\sqrt[3]{x-8} = 2\sqrt[3]{2x-6} \, .$$

解根号方程的第一步通常是将含变量的根号移到等号的一边，然后用乘方将根号转换成普通的多项式。我们用下面的例子来演示步骤。

To solve radical equations, the first step is usually isolate radical terms to one side of the equation, then raise its power such that radicals can be converted to normal polynomials. We demonstrate this using one example.

$$\sqrt{x+4} - x + 2 = 0$$
$$\sqrt{x+4} = x - 2$$
$$\left(\sqrt{x+4}\right)^2 = (x-2)^2$$
$$x + 4 = x^2 - 4x + 4$$
$$x^2 - 5x = 0$$
$$x(x-5) = 0$$
$$x = 0 \text{ and } x = 5$$

我们需要验证上述两个解:

$x = 0$	$x = 5$
$\sqrt{0+4} - 0 + 2 = 0$	$\sqrt{5+4} - 5 + 2 = 0$
$2 - 0 + 2 = 0$	$3 - 5 + 2 = 0$
$4 = 0$ 不成立	$0 = 0$ 成立
$x = 0$ 不是解	$x = 5$ 是解

求得的解可能不是真正的解的原因是这样的：当我们乘方时，得出的方程与原来的方程并不等同。

$$x = 1$$
$$x^2 = 1^2 = 1$$
$$x = 1 \text{ 和 } x = -1$$

在上面的例子里，平方 $x = 1$，错误地引进了不正确的解 $x = -1$。

We need to verify the two solutions:

$x = 0$

$\sqrt{0+4} - 0 + 2 = 0$

$2 - 0 + 2 = 0$

$4 = 0$ *is false*

$x = 0$ *is not a solution*

$x = 5$

$\sqrt{5+4} - 5 + 2 = 0$

$3 - 5 + 2 = 0$

$0 = 0$ *is true*

$x = 5$ *is a solution*

The reason that there might be false solution is because when we raise the power of an equation, the resulting equation is not identical to the original one.

$x = 1$

$x^2 = 1^2 = 1$

$x = 1$ and $x = -1$

In the example above, by square the equation $x = 1$ we erroneously generated another solution $x = -1$.

11 函数Functions

我们给出了很多函数的例子，但是还没有正式介绍函数的普适定义。

函数是一个将集合 X 中的元素与 Y 中的一个元素相匹配的运算，要求 X 中的每个元素在 Y 中都有且只有一个对应的元素。集合 X 称为定义域，集合 Y 称为对应域（到达域，陪域，上域）。例如，函数 *eye_color* 记录集合 X 中所有人的眼睛颜色。集合 Y 中列出了一个人可能的眼睛颜色。

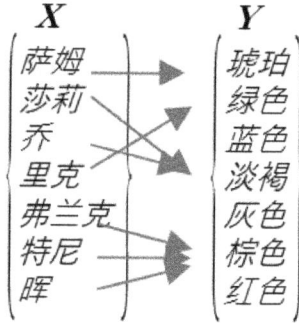

集合 X={*萨姆，莎莉，乔，里克，弗兰克，特尼，晖*}是定义域，集合 Y={ *琥珀，绿色，蓝色，淡褐，灰色，棕色，红色*}是对应域。请注意，X 中每个元素在 Y 中都有一个对应的元素，但反之则不然。Y 中的一些元素可能在 X 中具有多个对应元素。并且只有 Y 中的部分元素在 X 中具有对应元素，这些 Y 中的成员组成一个子集 *R = {琥珀，绿色，淡褐，棕色}*。子集 R 叫做值域。

We have given many examples of functions, but have not formally introduced what is a function in general.

Function is an operation that assigns an element in the set X to one element in Y, with the requirement that for each element in X there will be exactly one corresponding element in Y. Set X is called domain and set Y is called codomain.

For example, the *eye_color* function records a person's eye color for everyone in the set X. A person's possible eye colors are listed in set Y.

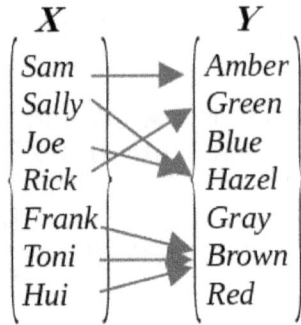

The set X={*Sam, Sally, Joe, Rick, Frank, Toni, Hui*} is the domain, and the set Y={*Amber, Green, Blue, Hazel, Gray, Brown, Red*} is the codomain. Note each element in X has one corresponding element in Y, but the reverse is not true. Some elements in Y may have more than one corresponding elements in X. Only some elements in Y have corresponding elements in X, which forms a subset R = {*Amber, Green, Hazel, Brown*}. Subset R is called range.

定义域和值域 Domain and Range

函数 $y = f(x)$ 的定义域是所有可能的 x 值，值域是相对应的所有输出 y。

例如函数 $3\sqrt{(2x+12)} - 5$ 中 x 值是受限制的，因为 $2x + 12$ 不能是负数。所以定义域是 $2x + 12 \geq 0$，或者 $x \geq -6$，也可以写成 $[-6, \infty)$ 或者 $\{x \mid x \geq -6\}$。因为 $\sqrt{} \geq 0$，所以函数的值域是 $y \geq -5$，写作 $[-5, \infty)$ 或者 $\{y \mid y \geq -5\}$。如果端点包含在集合中，则使用方括号；如果端点不在集合中，则使用小括号。

For function $y = f(x)$, the domain is all possible values of x, and the range is all output y.

For example, the x value for the function $3\sqrt{(2x+12)} - 5$ is restricted because $2x + 12$ must not be negative. So the domain of the function is $2x + 12 \geq 0$, or $x \geq -6$, also written as $[-6, \infty)$ or $\{x \mid x \geq -6\}$. Because $\sqrt{} \geq 0$, so the range of the function is $y \geq -5$, written as $[-5, \infty)$ or $\{y \mid y \geq -5\}$. Square bracket is used if endpoint is included in the set, parentheses is used if endpoint is not in the set.

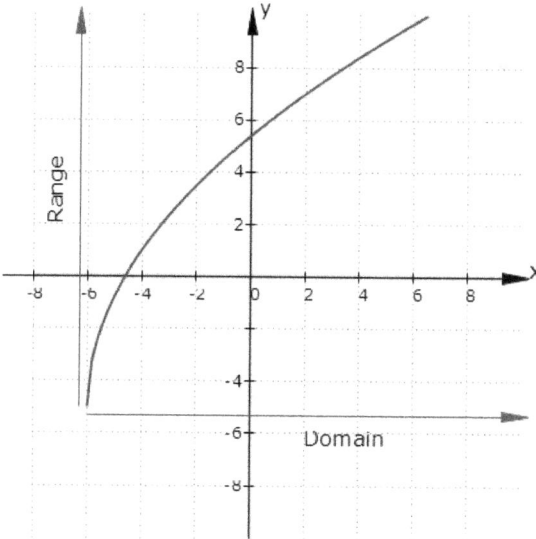

Figure 14: Domain and range of the radical function $3\sqrt{(2x+12)} - 5$. 根式函数 $3\sqrt{(2x+12)} - 5$的定义域和值域。

反函数 Inverse Function

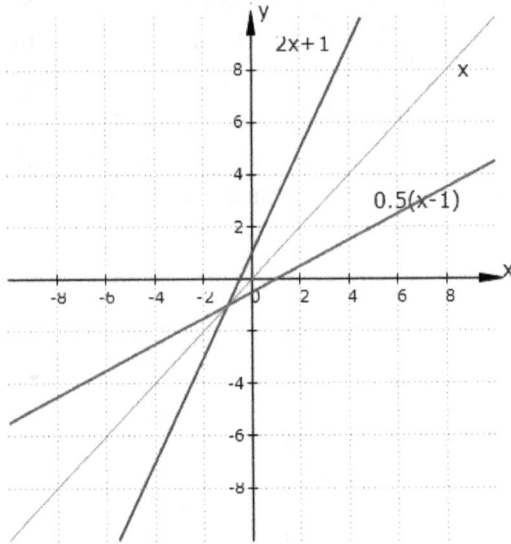

Figure 15: 2x + 1 and 0.5(x - 1）are inverse functions of each other. They are symmetric about the line y = x. 方程 2x + 1 和 0.5(x - 1) 互为逆函数，两者依 y = x 对称。

函数$f(x)(X \rightarrow Y)$的反函数（逆函数）为$f^{-1}(x)(Y \rightarrow X)$，它将每一个$y$值映射到一个唯一的$x$值，使得$f(x) = y$。

例如：

$$f(x) = 2x + 1, \, y = 2x + 1, \, x = \frac{1}{2}(y - 1),$$

互换x和y，我们得到

$$f^{-1}(x) = y = \frac{1}{2}(x - 1)。$$

For a function $f(x)(X \to Y)$, the inverse function $f^{-1}(x)(Y \to X)$ maps each element y to the unique element x such that $f(x) = y$.

Take the example:

$$f(x) = 2x + 1, \, y = 2x + 1, \, x = \frac{1}{2}(y - 1),$$

now swap x and y, we have

$$f^{-1}(x) = y = \frac{1}{2}(x - 1).$$

$f(x)$和$f^{-1}(x)$关于 $y = x$ 对称（图 15）。$f(x)$的定义域是$f^{-1}(x)$的值域；$f(x)$的值域是的$f^{-1}(x)$定义域。

$f(x)$ and $f^{-1}(x)$ are symmetric about the line $y = x$ (Figure 15). The domain for $f(x)$ is the range for $f^{-1}(x)$; the range for $f(x)$ is the domain for $f^{-1}(x)$.

首先$f^{-1}(x)$不应该混淆为$(f(x))^{-1}$。后者是乘法的逆运算（倒数），不是逆函数。

其次不是所有的函数都有逆函数。一个简单的例子是 $y = x^2$，对应于一个y值$y = 4$，有两个x值$x = \pm 2$。那么从y映射到x就不是一一对应，所以没有逆函数。解决这个困境的方法是限制定义域和值域。如果我们限制函数$f(x) = x^2$的定义域为$[0, \infty)$，那么逆函数$f^{-1}(x)$的值域必然也是$[0, \infty)$。也就是说$f^{-1}(x)$一定是非负值，因此$f^{-1}(x)$必须是$\sqrt{x}$，而不可能是$-\sqrt{x}$。

First $f^{-1}(x)$ should not be confused with $(f(x))^{-1}$. The later is multiplicative inverse (reciprocal), not an inverse function.

Second, not every function $f(x)$ has an inverse function. A simple example is $y = x^2$, for the value $y = 4$, there are two values $x = \pm 2$. So mapping back from y to x is not one-to-one; therefore, there is no unique inverse function. There are ways to get around this by restricting domain and range. For the function $f(x) = x^2$, if we restrict domain to $[\,0\,,\infty\,)$, then the range for the inverse function $f^{-1}(x)$ is also $[\,0\,,\infty\,)$. This implies $f^{-1}(x)$ must be non-negative, thus $f^{-1}(x)$ must be $\sqrt{x}$, ruling out $-\sqrt{x}$.

复合函数 Composition of Functions

两个函数，如果一个函数的结果被用作另一个函数的变量，这种运算叫做复合函数。比如两个函数 $f(x)$ 和 $g(x)$，根据哪个函数的结果被用作变量，我们会得到不同的复合函数 $f(g(x))$ 或者 $g(f(x))$。假如 $f(x) = 2x + 3$，$g(x) = 2x^2 + 2x - 1$，我们得到

Given two functions, one function's result can be passed as the argument for the other function. This operation is called composition of functions. For two functions $f(x)$ and $g(x)$, depending on which one is passed as the argument, the resulting function can be $f(g(x))$ or $g(f(x))$. For $f(x) = 2x + 3$, $g(x) = 2x^2 + 2x - 1$, we have

$$f\big(g(x)\big) = 2\,g(x) + 3 = 2\big(2x^2 + 2x - 1\big) + 3$$
$$= 4x^2 + 4x + 1,$$
$$g\big(f(x)\big) = 2\big(g(x)\big)^2 + 2\,g(x) - 1 = 2\big(2x+3\big)^2 + 2\big(2x+3\big) - 1$$
$$= 2\big(4x^2 + 12x + 9\big) + 4x + 6 - 1$$
$$= 8x^2 + 28x + 23.$$

函数 $f(x)$ 和逆函数 $f^{-1}(x)$ 的复合函数是

For function $f(x)$ and its inverse function $f^{-1}(x)$, their composition functions are

$$f\big(f^{-1}(x)\big) = x,$$
$$f^{-1}\big(f(x)\big) = x.$$

12 二次方程和复数 Quadratic Equations and Complex Numbers

虚数和复数 Imaginary and Complex Numbers

二次方程 $a x^2 + b x + c = 0$ 有下列解

$$x = \frac{-b \pm \sqrt{b^2 - 4ac}}{2a} \text{。}$$

但是当判别式 $\Delta = b^2 - 4ac < 0$ 时，平方根不是一个实数。我们需要引入一个全新的数，虚数。

首先让我们来解释虚数的单位 $i = \sqrt{-1}$ ，它有以下特性：

$i^0 = 1$ ，
$i^2 = -1$ ，
$i^3 = -i$ ，
$i^4 = 1$ 。

任何小于 0 的数 $x < 0,$

$$\sqrt{x} = i \sqrt{|x|}$$

是一个虚数。

二次方程的解可以写成

$$x = \frac{-b \pm i \sqrt{|b^2 - 4ac|}}{2a} \text{，如果 } b^2 - 4ac < 0 \text{。}$$

上面的解既有实数部分又有虚数部分，叫做复数。

The quadratic equation $a x^2 + b x + c = 0$ has the following solutions

$$x = \frac{-b \pm \sqrt{b^2 - 4\,ac}}{2\,a}\,.$$

However when the discriminant $\Delta = b^2 - 4\,ac < 0$, the square root value is not a real number. We must introduce a new number type called imaginary number.

First we introduce the imaginary number unit $i = \sqrt{-1}$. It has the following properties:

$i^0 = 1$,
$i^2 = -1$,
$i^3 = -i$,
$i^4 = 1.$

And for any $x < 0$,

$$\sqrt{x} = i\,\sqrt{|x|}$$

is called an imaginary number.

The solutions of the quadratic equation can then be written as

$$x = \frac{-b \pm i\,\sqrt{|b^2 - 4\,ac|}}{2\,a}\,, \text{ if } b^2 - 4\,ac < 0.$$

The above solutions are a mix of real numbers and imaginary numbers, which are called complex numbers.

复数的标准形式是 $a + b\,i$（a, b 为实数）。其算术运算满足以下规则：

$$(a + b\,i) + (c + d\,i) = a + c + (b + d)\,i\,,$$
$$\begin{aligned}(a + b\,i)(c + d\,i) &= ac + ad\,i + bc\,i + bd\,i^2\\ &= ac - bd + (ad + bc)\,i\,。\end{aligned}$$

运用虚数 i 时必须小心，避免下面的悖论：

$$-1 = i^2 = \sqrt{-1}\,\sqrt{-1} = \sqrt{(-1)(-1)} = \sqrt{1} = 1\,。$$

造成这个悖论的原因是 $\sqrt{-1}$ 不是普通的实数平方根。普通平方根的规则（例如 $\sqrt{x}\,\sqrt{y} = \sqrt{xy}$）不适用于负的 x 和 y。

另外，两个复数 $a+ib$ 和 $a-ib$ 称为共轭复数。二次方程的两个复数解一定是共轭复数。

The standard form of a complex number is $a + bi$ (a, b are real numbers). The rules of arithmetic operations are as follows:

$$(a+bi)+(c+di)=a+c+(b+d)i，$$
$$(a+bi)(c+di)=ac+adi+bci+bdi^2$$
$$=ac-bd+(ad+bc)i.$$

Care must be taken when we use imaginary unit because of the following fallacy:

$$-1=i^2=\sqrt{-1}\sqrt{-1}=\sqrt{(-1)(-1)}=\sqrt{1}=1.$$

The reason for this paradox is that $\sqrt{-1}$ is not an ordinary real-number square root. The rules for ordinary square root such as $\sqrt{x}\sqrt{y}=\sqrt{xy}$ don't hold for negative x and y.

Last note, two complex numbers $a+ib$ and $a-ib$ are called complex conjugate of each other. Complex solutions of quadratic equations are always complex conjugates.

二次方程复数解 Quadratic Equation with Complex Solutions

举一个简单的二次方程例子

Take a simple example of quadratic equation

$$x^2+16=0，$$
$$x^2=-16，$$
$$x=\pm\sqrt{-16}=\pm i\sqrt{16}=\pm 4i.$$

也可以使用相同的完成平方的方法。例如，

The same completing the square method can also be used. For example,

$$x^2 + 4\,x + 8 = 0 \, ,$$
$$x^2 + 4\,x + 4 + 4 = 0 \, ,$$
$$(x+2)^2 = -4 \, ,$$
$$x + 2 = \pm \sqrt{-4} = \pm 2\,i \, ,$$
$$x = -2 \pm 2\,i \, .$$

还可以使用求解二次方程的公式：

One can also use the formula for solving quadratic equation:

$$x^2 + 4\,x + 8 = 0 \, ,$$
$$x = \frac{-b \pm \sqrt{b^2 - 4\,ac}}{2\,a} = \frac{-4 \pm \sqrt{4^2 - 4 \cdot 1 \cdot 8}}{2 \cdot 1}$$
$$= \frac{-4 \pm \sqrt{-16}}{2} = \frac{-4 \pm 4\,i}{2} = -2 \pm 2\,i \, .$$

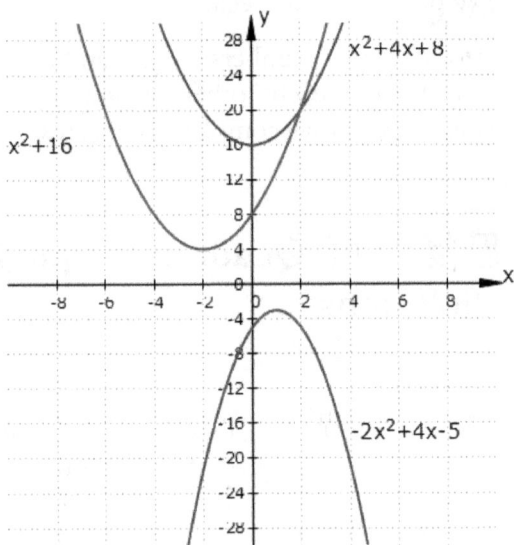

Figure 16: Examples of quadratic equations with complex roots. Their graphs don't intercept x-axis. 有复数根的二次方程例子。它们的曲线不与 x 轴相交。

13　多项式方程Polynomial Equations

一个变量的多项式方程具有标准形式

$$\sum_{k=0}^{n} a_k\, x^k = a_n\, x^n + a_{n-1}\, x^{n-1} + \cdots + a_2\, x^2 + a_1\, x + a_0 = 0 \text{,}$$

这里 $a_n \neq 0$。这个方程也称为单变量多项式方程。

The polynomial equation in one indeterminate has the standard form

$$\sum_{k=0}^{n} a_k\, x^k = a_n\, x^n + a_{n-1}\, x^{n-1} + \cdots + a_2\, x^2 + a_1\, x + a_0 = 0 \text{,}$$

where $a_n \neq 0$. It is also called univariate polynomial equation.

因子分解法 Solving By Factoring

简单的多项式方程可以通过试探和因式分解来求解。举个例子

Simple polynomial equations can be solved by inspection and factoring. Take an example

$$x^4 - x = 0.$$

通过试探，我们可以看到 0 和 1 是方程的两个根，因此 x 和 $x-1$ 是多项式的因子。然后很容易将方程重写为

By inspection, we can see 0 and 1 are two roots of the equation, so x and $x-1$ are factors of the polynomial. It is then easy to rewrite the equation as

$$x\left(x^3 - 1\right) = 0$$
$$x\left(x^3 - x^2 + x^2 - x + x - 1\right) = 0$$
$$x\left(x - 1\right)\left(x^2 + x + 1\right) = 0$$
$$x = 0,\ x - 1 = 0,\ x^2 + x + 1 = 0.$$
$$x = 0,\ x = 1,\ x = \frac{-1 \pm i\sqrt{3}}{2}.$$

该方程有 4 个根，恰好是多项式中的最高指数，或多项式的次数。每个 n 次单变量多项式 ($n \geq 1$) 都可以分解为

$$a_n\left(x - r_1\right) \cdots \left(x - r_n\right),$$

这里 $a_n, r_1, \cdots, r_n$ 是复数，并且多项式方程有 n 个根 $r_1, \cdots, r_n$。这称为 n 次方程有 n 根定理，它可以从代数基本定理推导出来。

The equation has 4 roots, exactly the highest exponent in the polynomial, or the degree of the polynomial. Every univariate polynomial of degree n ($n \geq 1$) can be factorized as

$$a_n\left(x - r_1\right) \cdots \left(x - r_n\right),$$

where $a_n, r_1, \cdots, r_n$ are complex numbers, and there are n roots $r_1, \cdots, r_n$ to the polynomial equation. This is called n-roots theorem, which can be derived from the fundamental theorem of Algebra.

代数基本定理：每个具有复系数的非常数单变量多项式都至少有一个复数根。

由于实数和整数都是复数的特例，因此 n 次方程有 n 根定理和代数基本定理对于具有实数或整数系数的多项式同样成立。

The fundamental theorem of Algebra: Every non-constant univariate polynomial with complex coefficients has at least one complex root.

Since real-numbers and integers are all special cases of complex numbers, the n-roots theorem and the fundamental theorem of Algebra are also true for polynomials with real-number or integer coefficients.

图解法 Solving By Graphing

图示有助于求解多项式方程。

（1）通过作图，我们可以判断方程是否有实数根。如果曲线与 x 轴相交，则将存在实数根（图 17，$P(x)$、$Q(x)$ 和 $R(x)$）；如果图形与 x 轴不相交，则不会有实数根（图 17，$S(x)$）。

（2）曲线与 x 轴相交的次数表明实数根的个数。图 17，$P(x)$ 和 $Q(x)$ 有一个实数根，$R(x)$ 有三个实数根。

（3）假如多项式有一个形式为 $(x-a)^k$ 的因子，这里 a 是一个实数，k 是一个整数，那么曲线将在 $x=a$ 的点于 x 轴相交。如果 k 是偶数，曲线只会与 x 轴相切（图 17，$P(x)$）；如果 k 是奇数，曲线将穿过 x 轴（图 17，$Q(x)$）。

（4）一个方程 $f(x)$，如果 $x1$ 和 $x2$ 是实数，而且 $f(x1)$ 和 $f(x2)$ 数值符号相反，在 $x1$ 和 $x2$ 之间一定有一个根。例如图 17 $R(x)$, $R(-2)=2$, $R(6)=-6$, 两者符号相反。所以 -2 和 6 之间一定有一个实数根。事实上 $R(x)$ 有三个根。

Graphing can be helpful in solving polynomial equations.

(1) By graphing, we may determine whether the equation has real-number roots. If the graph intercepts the x-axis, then there will be real-number roots (Figure 17, $P(x)$, $Q(x)$ and $R(x)$); if the graph does not intercept the x-axis, then there will be no real-number roots (Figure 17, $S(x)$).

(2) The number of times the graph intercepts the x-axis suggests the number of real-number roots. In Figure 17, $P(x)$ and $Q(x)$ has 1 real-number roots, $R(x)$ has 3 real roots.

(3) Polynomial may have a factor in the form of $(x-a)^k$, where a is a real number and k is a positive integer. The graph

will intercept the x-axis at $x = a$. If k is even, the graph will only touch the x-axis (Figure 17, *P(x)*); if k is odd, the graph will cross the x-axis (Figure 17, *Q(x)*).

(4) For a function *f(x)*, if *x1* and *x2* are real numbers, *f(x1)* and *f(x2)* has opposite signs, there must exist a root between *x1* and *x2*. For example, Figure 17 *R(x)*, *R(−2)=2, R(6)=−6,* they have opposite signs. There must be at least one real-number solution between −2 and 6. Indeed *R(x)* has three roots.

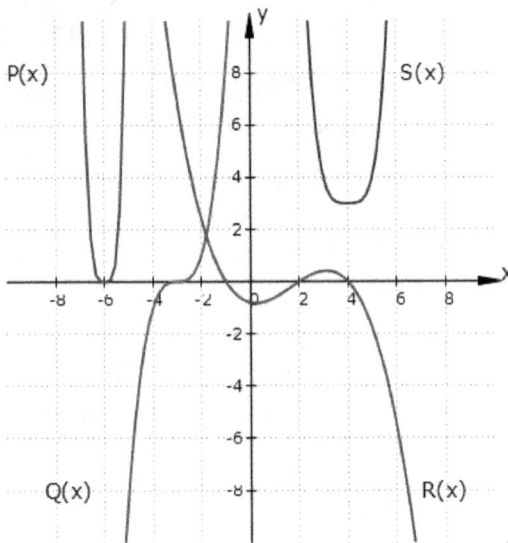

Figure 17: Examples of polynomial equations. S(x) has 0 root, P(x) and Q(x) have 1 root, R(x) has 3 roots.多项式方程的例子。S(x)没有根，P(x)和Q(x)有一个根，S(x)有三个根。

69

笛卡尔符号规则 Descartes' rule of signs

笛卡尔符号规则给出了具有实数系数的多项式方程的正或负（非零）实数根的上限。规则如下：

（1）数一数多项式方程$f(x)(a_0 \neq 0)$相邻系数之间的符号变化次数。正实根的个数等于符号变化次数减去 2 的倍数。最大为符号变化次数，最小为 0。

（2）数一数多项式方程$f(-x)(a_0 \neq 0)$相邻系数之间的符号变化次数。负实根的个数等于符号变化次数减去 2 的倍数。最大为符号变化次数，最小为 0。

例如：

$$f(x) = 2x^3 - x^2 + 6x - 3$$

f(x) 有三次符号改变(2 到–1，–1 到 6，6 到–3)，所以方程可以有 3、1 或者 0 个正实数根。

$$f(-x) = -2x^3 - x^2 - 6x - 3$$

f(–x) 没有符号变化，所以 f(x) 有 0 个负实数根(f(–x) 有 0 个正实数根)。

Descartes' rule of signs gives an upper bound of how many positive or negative (non-zero) real roots of a polynomial equation with real coefficients. The rules are as following:

(1) For $f(x)(a_0 \neq 0)$, count the number of sign changes between consecutive coefficients. The number of positive real roots equals the count of sign changes minus multiples of 2. The maximum is the counts of sign changes, the minimum is 0.

(2) For $f(-x)(a_0 \neq 0)$, count the number of sign changes between consecutive coefficients. The number of negative real

roots equals the count of sign changes minus multiples of 2. The maximum is the counts of sign changes, the minimum is 0.

For example:

$$f(x) = 2x^3 - x^2 + 6x - 3$$

$f(x)$ has 3 sign changes (2 to -1, -1 to 6, 6 to -3), so it may have 3, 1 or 0 positive real roots.

$$f(-x) = -2x^3 - x^2 - 6x - 3$$

$f(-x)$ has 0 sign changes, so $f(x)$ has 0 negative real root ($f(-x)$ has 0 positive real root).

有理根定理Rational Root Theorem

有理根定理: 对于具有整数系数的单变量多项式，如果存在任何有理根 p/q，则 p 必须是 a_0 的整数因子，q 必须是 a_n 的整数因子。

可能的有理数根 $\dfrac{p}{q} = \pm \dfrac{a_0 \text{ 的整数因子}}{a_n \text{ 的整数因子}}$

在我们使用有理根定理求多项式方程的有理根之前，我们应该澄清并非所有多项式方程都有有理根。举个简单的例子，$x^2 - 2 = 0$，它的根 $x = \pm\sqrt{2}$，是无理数。

例子：

$$f(x) = 2x^3 - x^2 + 6x - 3 = 0 \text{，}$$
$$a_0 = 3 \text{，} \quad a_n = 2 \text{。}$$

a_0 的整数因子是 1 和 3，a_n 的整数因子是 1 和 2。考虑到所有可能的组合，我们有

可能的有理数根 $\dfrac{p}{q} = \pm\dfrac{1}{1}, \pm\dfrac{3}{1}, \pm\dfrac{1}{2}, \pm\dfrac{3}{2}$。

运用笛卡尔符号规则，我们发现方程 *f(x)* 有 0 个负根，因此我们只需要考虑正数。——验证，我们会发现 ½ 是唯一的一个有理根。½ 是 *2x − 1 = 0* 的根。我们可以通过分解进一步求解方程，如下所示：

$$2\,x^3 - x^2 + 6\,x - 3 = 0 \text{，}$$
$$x^2(2\,x - 1) + 3(2\,x - 1) = 0 \text{，}$$
$$(2\,x - 1)(x^2 + 3) = 0 \text{。}$$
$$2\,x - 1 = 0 \text{，} x^2 + 3 = 0 \text{。}$$
$$x = \frac{1}{2} \text{，} x = \pm\,i\,\sqrt{3} \text{。}$$

Rational root theorem: For a univariate polynomial with integer coefficients, if there exists any rational root p/q, then p must be an integer factor of a_0, q must be an integer factor of a_n.

$$\text{possible rational roots } \frac{p}{q} = \pm\,\frac{\text{integer factor of } a_0}{\text{integer factor of } a_n}$$

Before we use the rational root theorem to find rational roots of polynomial equations, we should clarify that not all polynomial equations have rational roots. Take the simple example $x^2 - 2 = 0$, its roots are $x = \pm\,\sqrt{2}$, which are irrational numbers.

Example:

$$f(x) = 2\,x^3 - x^2 + 6\,x - 3 = 0 \text{,}$$
$$a_0 = 3 \text{, } a_n = 2.$$

The integer factors of a_0 are 1 and 3, the integer factors of a_n are 1 and 2. Considering all possible combinations, we have

$$\text{possible rational roots } \frac{p}{q} = \pm\,\frac{1}{1}\,, \pm\,\frac{3}{1}\,, \pm\,\frac{1}{2}\,, \pm\,\frac{3}{2}\,.$$

Using Descartes' rule of signs, we find the equation *f(x)* has 0 negative roots, so we need to consider only positive numbers. Verifying one by one, we'll find ½ is the only rational root. ½ is the root for *2x − 1 = 0*. We can further solve the equation by factorization as following:

$$2x^3 - x^2 + 6x - 3 \quad = 0,$$
$$x^2(2x - 1) + 3(2x - 1) = 0,$$
$$(2x - 1)(x^2 + 3) \quad = 0.$$
$$2x - 1 = 0, \quad x^2 + 3 = 0.$$
$$x = \frac{1}{2}, \quad x = \pm i\sqrt{3}.$$

无 理 共 轭 定 理 The Irrational Conjugate Theorem

无理共轭定理：如果 $a + \sqrt{b}$ 是有理系数多项式的一个无理根，则其无理共轭 $a - \sqrt{b}$ 也是一个根。这里 $a\ b$ 都是有理数并且 $\sqrt{b}$ 是无理数。例如，如果 1 和 $3 + \sqrt{2}$ 是一个方程的两个根，那么 $3 - \sqrt{2}$ 也是方程的一个根。原方程应该是

The irrational conjugates theorem：that if $a + \sqrt{b}$ is an irrational root of a polynomial with rational coefficients, then its irrational conjugate $a - \sqrt{b}$ is also a root. Here $a\ b$ are both rational numbers and $\sqrt{b}$ is irrational.

For example if an equation has roots 1 and $3 + \sqrt{2}$, then it must also have root $3 - \sqrt{2}$. The original equation must be

$$(x - 1)(x - (3 - \sqrt{2}))(x - (3 + \sqrt{2}))$$
$$= (x - 1)(x - 3 + \sqrt{2})(x - 3 - \sqrt{2})$$
$$= (x - 1)((x - 3)^2 - 2).$$

14 指 数 和 对 数 Exponential and Logarithm

指数函数Exponential Functions

指数函数的形式为

$$f(x) = a^x, \quad a > 0 \text{。}$$

$f(0) = a^0 = 1$ 对于所有正实数 a。
$f(x) > 0$ 对于所有实数 x。
$f(x+y) = a^{x+y} = a^x a^y$
$\quad = f(x) f(y)$ 对于所有实数 x 和 y。

下面的函数也会被称为指数函数：

$$f(x) = \exp(x) = e^x \text{。}$$

$e = 2.71828 \cdots$ 是一个无理数常数，通常叫常数 e、自然基数或者欧拉数。

一个计算 e 的数值的方法是用一个大的 n 数值来计算 $\left(1+\dfrac{1}{n}\right)^n$，当 n 趋近无穷大时，$\left(1+\dfrac{1}{n}\right)^n$ 趋近 e。e 在数学中经常自然出现，上述极限就是一个例子。我们将在第 17 章看到另一个例子，欧拉方程。

Exponential functions are of the form

$$f(x) = a^x, \quad a > 0.$$

$f(0) = a^0 = 1$ *for all positive real values of a* .
$f(x) > 0$ *for all real values of x* .
$f(x+y) = a^{x+y} = a^x a^y$
$\quad = f(x) f(y)$ *for all real values of x and y* .

Exponential function can sometimes refer to

$$f(x) = \exp(x) = e^x ,$$

where $e = 2.71828 \cdots$ is an irrational constant, often called the number e, natural base, also known as Euler's number.

One method to calculate the value e is to calculate the value $\left(1 + \dfrac{1}{n}\right)^n$ for large integer n, since as n approaches infinity, $\left(1 + \dfrac{1}{n}\right)^n$ approaches e. The number e arises naturally in mathematics frequently, and the above limit is one example. We'll see another example in Chapter 17, Euler's equation.

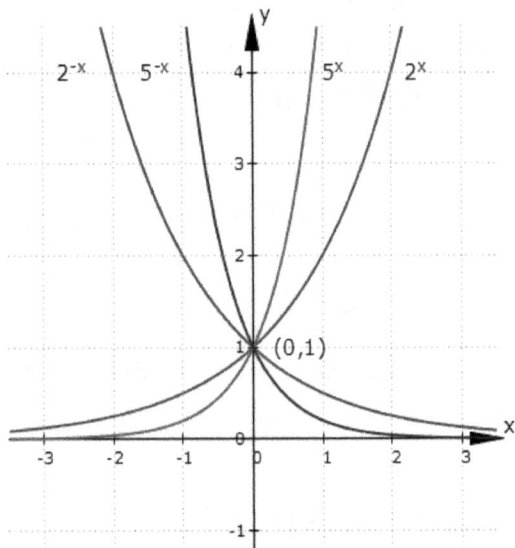

Figure 18: Graphs of exponential functions. All cross at (0, 1). a^x and a^{-x} are symmetric about y-axis. 指数函数图像都通过(0, 1)。a^x 和 a^{-x} 依 y 轴对称。

在上图中（图18）我们画出了几个指数函数的图形 $(a>1)$。如果 $a=1$，$f(a,x)=1^x=1$。如果 $a<1$，a^x 的图形与 $\left(\dfrac{1}{a}\right)^{-x}$ 相同，因为 $a^x=\left(\dfrac{1}{a}\right)^{-x}$。

In the figure above (Figure 18) we graphed several exponential functions with $a>1$. For $a=1$, $f(a,x)=1^x=1$. For $a<1$, the shape of a^x is identical to $\left(\dfrac{1}{a}\right)^{-x}$, because $a^x=\left(\dfrac{1}{a}\right)^{-x}$.

指数增长和衰减 Exponential Growth and Decay

复利 复利是指数增长的一个例子。假如莎拉在银行储蓄账户中存入 4000 美元，利息每月复利，年利率为 7%。两年后储蓄账户里会有多少钱？

复利计算公式如下：

$$A = P\left(1+\frac{r}{n}\right)^{nt}，$$

这里 P 是最初的存款，r 是年利率，n 是每年复利计算的次数，t 是存款年数，A 是最终账号里的存款数目。用计算公式，我们算得

$$A = 4000\left(1+\frac{0.07}{12}\right)^{12\cdot 2}$$
$$= 4599.22$$

Compound Interest Compound interest is an example of exponential growth. Suppose Sarah invests $4000 in a bank's saving account. The interest in compounded every month, with an annual interest rate of 7%. How much is in the saving account after two years?

The formula for calculate compound interest is as follows:

$$A = P \left(1 + \frac{r}{n}\right)^{nt},$$

where P is the initial principal amount, r is the annual interest rate, n is the number of times the interest rate is compounded annually, t is time period in years, and A is amount at the end of time period. Using the above formula, we find

$$A = 4000 \left(1 + \frac{0.07}{12}\right)^{12 \cdot 2}$$
$$= 4599.22$$

半衰期 放射性衰变是指数衰变的一个例子。例如，碳 14 是一个放射性同位素，它会衰变为稳定的非放射性同位素氮 14。碳 14 的半衰期 $t_{1/2}$ 为 5700 年。10000 年后，剩下的 C-14 的百分比是多少？

放射性衰变的计算公式是

$$N = N_0 \left(\frac{1}{2}\right)^{\frac{t}{t_{1/2}}},$$

这里 N_0 是最初的放射性，t 是过去的时间。对于我们的例子我们得到

$$\frac{N}{N_0} = \left(\frac{1}{2}\right)^{10000/5700} = 0.5^{1.754} = 0.30 = 30\,\% \, 。$$

Half Life Radioactive decay is an example of exponential decay. For example Carbon-14 is a radioactive isotope which decays into a stable non-radioactive isotope Nitrogen-14. Carbon-14 has a half life $t_{1/2}$ of 5700 years. What percentage of C-14 is still present after 10000 years?

The formula for calculating radioactive decay is

$$N = N_0 \left(\frac{1}{2}\right)^{\frac{t}{t_{1/2}}},$$

where N_0 is the initial radioactivity, t is time passed. For our example, we have

$$\frac{N}{N_0} = \left(\frac{1}{2}\right)^{10000/5700} = 0.5^{1.754} = 0.30 = 30\,\% \,.$$

对数函数Logarithmic Functions

对数(log)是指数函数的反函数（图 19）。也就是说

$a^y = x$ ，
$y = \log_a(x) = \log_a x$ ，

a 是正实数，叫做基数。例如 $\log_2(8) = \log_2(2^3) = 3$ 。

如果基数是自然基数 e ，对数函数写成

$\ln x = \log_e x$ ，

叫做自然对数。

如果基数是10 , $\log_{10} x$ 叫做十进制对数或者常用对数，有时写成 $lg x$ 。

Logarithm is the inverse function to exponentiation (Figure 19). That is

$a^y = x$,
$y = \log_a(x) = \log_a x$,

where a is a positive real number, called the base. For example, $\log_2(8) = \log_2(2^3) = 3$.

If the base is the natural base e, the log function is usually written as

$\ln x = \log_e x$,

and it is called natural logarithm.

If the base is 10, $\log_{10} x$ is called decimal logarithm or common logarithm, sometimes written as *lg x*.

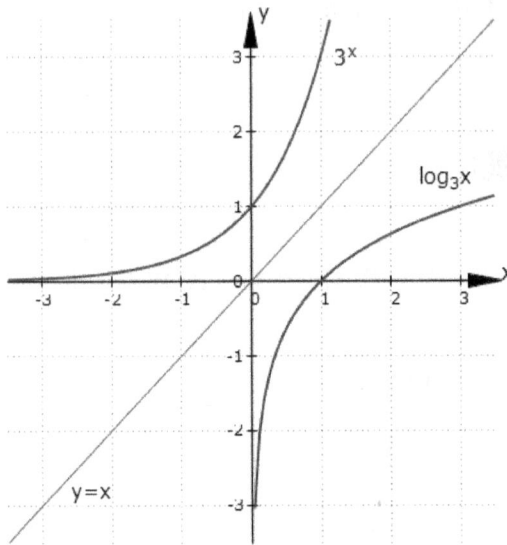

Figure 19: The exponential function 3^x and the logarithm function log₃x are inverse function of each other. 指数函数 3^x 和对数函数 log₃x 互为逆函数。

对数函数具有下列性质：

乘法性质 $\log_a (xy) = \log_a x + \log_a y$

除法性质 $\log_a \left(\dfrac{x}{y}\right) = \log_a x - \log_a y$

乘方性质 $\log_a (x^p) = p \log x$

开方性质 $\log_a \sqrt[p]{x} = \dfrac{\log_a x}{p}$

Logarithmic functions have the following properties:

$$\text{Product Property} \quad \log_a(xy) = \log_a x + \log_a y$$

$$\text{Quotient Property} \quad \log_a\left(\frac{x}{y}\right) = \log_a x - \log_a y$$

$$\text{Power Property} \quad \log_a(x^p) = p \log x$$

$$\text{Root Property} \quad \log_a \sqrt[p]{x} = \frac{\log_a x}{p}$$

对数换底公式 Change of Base Formula

对任何正实数基数 b，   For any positive real base b,

$$\log_a x = \frac{\log_b x}{\log_b a} \quad \circ$$

这个公式常常用来将计 算转换为以 10 或者自然数 e 为基数的运算。

This formula is frequently used to convert calculations using base 10 or natural number e.

$$\log_a x = \frac{\log_{10} x}{\log_{10} a} = \frac{\ln x}{\ln a} .$$

对数尺度 Log-Scale

当一个变量的数值变化范围很大时，用对数刻度绘制图表更 好。即使用 $\log_{10}(x)$ 值或 $\log_{10}(y)$ 值来代替原始数据。一个例子 是溶液的 pH 值，

$$pH = -\log_{10}\left[H^+\right] ,$$

这里 $\left[H^+\right]$ 是溶液中氢原子的摩尔浓度(mole/L)。当 x 轴为对 数尺度时，显示氢原子浓度和 pH 值关系的曲线是一条直线

（图 20）。注意虽然 x 轴用的是 *log* 数值，标注仍然用原来的数值。

When variable values change over a wide range, it is better to graph with log-scale. That is the values $\log_{10}(x)$ or $\log_{10}(y)$ are used instead of the original values. One example is the pH value of a solution,

$$pH = -\log_{10}\left[H^+\right],$$

where $\left[H^+\right]$ is the molar concentration (mole/L) of the hydrogen ion in a solution. The graph showing relationship between the pH value and hydrogen ion concentration shows a straight line when the x-axis is in *log*-scale (Figure 20). Note that although *log*-value is used for the x-variable, x-axis is still labeled in the original quantities.

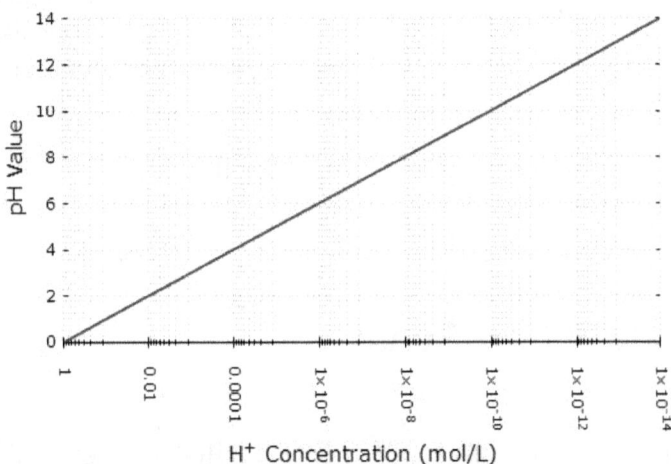

Figure 20: Relationship between H⁺ concentration and pH value. H⁺ concentration in log-scale. H⁺浓度和 pH 值的关系，H⁺浓度为对数尺度。

另一个例子是计算机芯片技术的指数增长。计算机的核心是一个称为芯片的微小部件，衡量其计算能力的指标之一是每个芯片中有多少个晶体管。图 21 显示了摩尔定律，即芯片中封装的晶体管数量呈指数增长。这个数字从 1971 年的几千增长到 2023 年的超过 1000 亿。通过在 y 轴上用对数刻度，随时间的指数增长近似为一条直线。

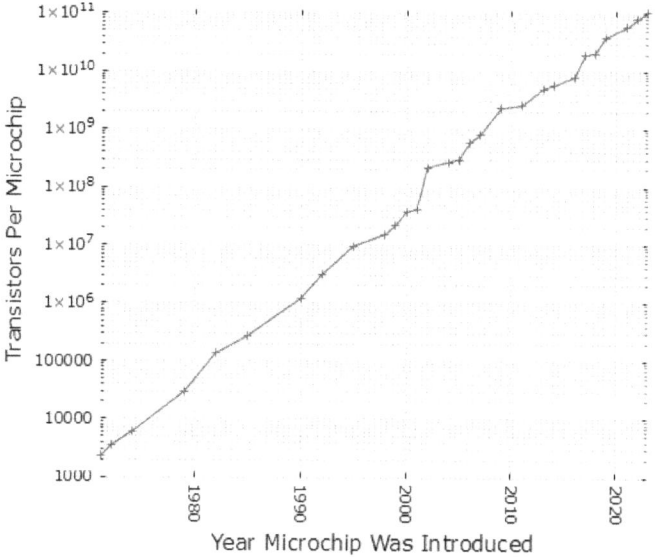

Figure 21: Moor's law shows the annual growth of transistors per microchip. y-axis in log-scale. 摩尔定律显示单个芯片中晶体管数目的逐年增长。晶体管数目使用对数尺度。

Another example is the exponential growth in computer chip technology. At the heart of a computer is a small device called microchip. One of the measures for its computational power is how many transistors in each microchip. The graph shows the Moor's law, namely the exponential grow of number of transistors packed in microchips. This number grew from a few thousands in 1971 to over 100 billions by 2023. By using *log*-scale on the *y*-axis, the exponential growth is showing as approximately a straight line over time.

指数和对数方程 Exponential and Logarithmic Equations

指数方程是指数中出现变量的方程。解指数方程的方法是将指数的基数变成同样的，然后利用下面的指数性质：

Exponential equations are equations in which variables appear as exponents. To solve exponential equations, we try to make the base the same and then make use of the following property of exponents:

$$a^x = a^y \rightarrow x = y \,.$$

如果基数不同，我们可以用对数换底：

If bases are not the same, we can use logarithm to change base:

$$a^x = b^y \,, \log_c a^x = \log_c b^y \,,$$

$$x\log_c a = y \log_c b \rightarrow x = \frac{\log_c b}{\log_c a} \, y \,.$$

最常用的基数（c）是自然基数 e，也就是

The base (c) for the *log* is frequently natural number e, i.e.,

$$x = \frac{\ln b}{\ln a} \, y \,.$$

例子 Example:

$$2^{x-1} = 3^{x+1} \,,$$
$$(x-1)\ln 2 = (x+1)\ln 3 \,,$$
$$x(\ln 3 - \ln 2) = -(\ln 2 + \ln 3) \,,$$
$$x = -\frac{\ln 3 + \ln 2}{\ln 3 - \ln 2} \,.$$

变量出现在对数中的方程叫对数方程。如果基数相同，如

Logarithmic equations are equations in which variables occur in logarithms. If the base is the same, for example,

$$\log_a x = \log_a y \rightarrow x = y .$$

如果基数不相同，我们可以改变基数。

If the bases are not the same, we can change the base.

$$\log_a x = \log_b y ,$$
$$\frac{\ln x}{\ln a} = \frac{\ln y}{\ln b}$$
$$\ln x = \frac{\ln a}{\ln b} \ln y ,$$
$$\ln x = \ln y^{\frac{\ln a}{\ln b}} ,$$
$$x = y^{\frac{\ln a}{\ln b}} .$$

例子 Example:

$$\log_2 (x - 1) = \log_4 (x + 5) ,$$
$$\frac{\ln (x - 1)}{\ln 2} = \frac{\ln (x + 5)}{\ln 4} ,$$
$$\ln 4 * \ln (x - 1) = \ln 2 * \ln (x + 5) ,$$
$$2 \ln (x - 1) = \ln (x + 5) ,$$
$$(x - 1)^2 = x + 5 ,$$
$$x^2 - 3x - 4 = 0 ,$$
$$(x - 4)(x + 1) = 0 ,$$
$$x = 4 \ \text{and} \ x = -1.$$

我们可以验证4是原方程的解，但−1不是。原因是如果 $x = -1$，则 $x - 1 = -2$ 将是一个负数，那么 $\log_2 (x - 1)$ 不是实数，但 $\log_4 (x + 5)$ 是实数。

One can verify that *4* is a solution to the original equation, however *−1* is not. Since if $x = -1$, $x - 1 = -2$ will be negative, thus $\log_2(x - 1)$ is not a real number, but $\log_4(x + 5)$ still is a real number.

15 有理方程Rational Equations

绘 制 有 理 函 数 图 Graphing Rational Functions

有理数函数具有下列形式

$$f(x) = \frac{P(x)}{Q(x)} \quad .$$

这里 $P(x)$和 $Q(x)$是多项式。多项式函数是有理函数的特例，其 $Q(x)$ 是一个常数。

有理表达式的运算规则与分数运算相似。比如要计算

$$\frac{x}{x+2} + \frac{2}{x-2} \quad ,$$

我们首先要找到分母的最小公倍数 LCM = $(x + 2)(x - 2)$。然后做以下运算：

$$\frac{x(x-2)}{(x+2)(x-2)} + \frac{2(x+2)}{(x-2)(x+2)}$$
$$= \frac{x(x-2) + 2(x+2)}{(x+2)(x-2)}$$
$$= \frac{x^2+4}{x^2-4} \quad 。$$

A rational function is of the form

$$f(x) = \frac{P(x)}{Q(x)} ,$$

where $P(x)$ and $Q(x)$ are polynomials. A polynomial function is a special case of rational function where $Q(x)$ is a constant.

Rules for arithmetic operations of rational expressions are similar to that for fractions. For example to calculate

$$\frac{x}{x+2} + \frac{2}{x-2} ,$$

we first find the LCM of denominators, which is *(x + 2)(x − 2)*. Then we have

$$\frac{x(x-2)}{(x+2)(x-2)} + \frac{2(x+2)}{(x-2)(x+2)}$$
$$= \frac{x(x-2) + 2(x+2)}{(x+2)(x-2)}$$
$$= \frac{x^2+4}{x^2-4} .$$

在图 22 中，我们绘出有理函数

$$y = \frac{2x+3}{x-1} 。$$

图形有两个分支。函数的定义域是 $x \neq 1$。如果 $x = 0$，$y = -3$，一个分支有 y-截距。如果 $y = 0$，$x = -3/2$，一个分支有 x-截距。当 $x \to \pm\infty$，$y \to 2$，曲线趋近一条水平线，叫做水平渐近线。当 $y \to \pm\infty$，$x \to 1$，曲线趋近一条垂直线，叫做垂直渐近线。

In Figure 22, we plotted the rational function

$$y = \frac{2x+3}{x-1} .$$

Note the graph have two branches. The domain of the function is *x ≠ 1*. If $x = 0$, $y = -3$, one branch has *y*-intercept. If $y = 0$, $x = -3/2$, one branch has *x*-intercept. As $x \to \pm\infty$, $y \to 2$, showing in the graph as a horizontal line called horizontal asymptote. As $y \to \pm\infty$, $x \to 1$, showing in the graph as a vertical line called vertical asymptote.

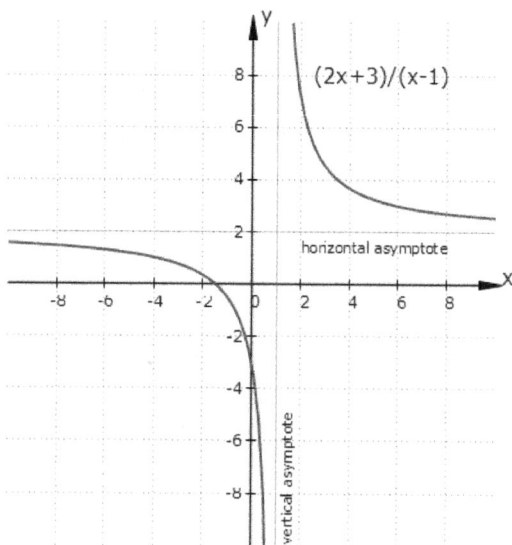

Figure 22: Graph of rational function
$y = \dfrac{2x+3}{x-1}$. 有理函数 $y = \dfrac{2x+3}{x-1}$ 的图形。

解有理方程 Solving Rational Equations

有理方程含有有理表达式 $\dfrac{P(x)}{Q(x)}$，这里 *P(x)* 和 *Q(x)* 是多项式。多项式方程是

A rational equation is an equation that contains rational expression $\dfrac{P(x)}{Q(x)}$, where *P(x)*

有理方程的特例。下面是一些有理方程的例子：

and $Q(x)$ are polynomials. Polynomial equation is a special case of rational equations. Here are some examples of rational equations:

$$x^3 - 1 = 0 \, ,$$
$$\frac{x - 1}{x^2 + x + 2} = \frac{1}{x} \, ,$$
$$\frac{1}{x - 1} - \frac{1}{x + 1} = 1.$$

解有理方程的方法是将方程两边乘以公分母，将分母消除。下面的例子显示了步骤：

To solve rational equations, we try to eliminate denominators by multiplying both sides by the least common multiple (LCM) of denominators. The steps are explained in the following example:

$$\frac{x - 1}{x^2 + x + 2} = \frac{1}{x} \, ,$$
$$LCD = x \left(x^2 + x + 2 \right) \, ,$$
$$\frac{x \left(x - 1 \right)}{x \left(x^2 + x + 2 \right)} = \frac{x^2 + x + 2}{x \left(x^2 + x + 2 \right)} \, ,$$
$$x \left(x - 1 \right) = x^2 + x + 2 \, ,$$
$$-2x = 2 \, , \, x = -1.$$

16 序列和系列Sequences and Series

序列（数列）是遵循某种规律的一组元素的集合。序列中元素的顺序很重要，通常写为

$a_1 , a_2 , a_3 , \ldots\ldots$

每个元素称为一个项，用下标1、2、3、……表示该项的位置。

A sequence is a set of elements that follow a regular pattern. The order of elements in the sequence is important, frequently written as

$a_1 , a_2 , a_3 , \ldots\ldots$

Each element is called a term, subscripts 1, 2, 3, ……denote the positions of the terms.

系列（级数）是一个序列中所有元素的和。如果系列的项数 N 是有限的，系列叫做有限系列，序列也叫做有限序列。有限系列写作

A series is the sum of all elements in a sequence. If the sequence has a finite N terms, the sequence is called a finite sequence, and the series is called a finite series. A finite series is written as

$$S_N = a_1 + a_2 + a_3 + \cdots + a_N = \sum_{k=1}^{N} a_k .$$

如果序列有无穷项，则叫做无限序列，相对应的系列叫做无限系列，写成

If the sequence has infinite terms, then it is called an infinite sequence, and the corresponding series is called an infinite series, written as

$$S = a_1 + a_2 + a_3 + \cdots = \sum_{i=1}^{\infty} a_i .$$

等差数列 Arithmetic Sequences

等差数列或等差序列是一个相邻的项相差一个常数 d 的序列：

An arithmetic sequence or arithmetic progression is a sequence where the successive terms are different by a constant value d:

$$a_{n+1} - a_n = d .$$

如果第一项是 a_1，序列的第 n 项是

If the initial term is a_1, the n-th term of the sequence is

$$a_n = a_1 + (n - 1) d .$$

下面的例子全是等差数列：

For example, the following are all arithmetic sequences:

$0 , 1 , 2 , 3 , 4 , \cdots$
$1 , 4 , 7 , 10 , \cdots$
$5 , 10 , 15 , 20 , 25$
$100 , 90 , 80 , 70 , 60 , 50$

与一个有限的等差数列相对应的系列是

For a finite arithmetic sequence, the corresponding series is

$$S_N = \sum_{i=1}^{N} a_i = \frac{N}{2} (a_1 + a_N)$$
$$= \frac{N}{2} (2 a_1 + (N - 1) d) .$$

将系列的项反向排列，与原来的系列相加，可以得到以上结果。

This can be proving by rewriting the terms in reverse order and adding them to the original sequence.

$$S_N = a_1 + a_2 + \cdots + a_{N-1} + a_N$$
$$+ S_N = a_N + a_{N-1} + \cdots + a_2 + a_1$$
$$2\,S_N = (a_1 + a_N) + \cdots + (a_1 + a_N) = N\,(a_1 + a_N)$$
$$S_N = \frac{N}{2}\,(a_1 + a_N).$$

用以上公式，我们可以计算整数 1 到 100 的和：

Using the above formula we can calculate the sum of all integers from 1 to 100:

$$S_{100} = \frac{100}{2}\,(1 + 100) = 5050.$$

如等差数列是无限的，$S_{N \to \infty} = \pm\infty$ 。

For an infinite arithmetic series, $S_{N \to \infty} = \pm\infty$.

等比数列 Geometric Sequences

等比数列的每一项是前一项乘以一个非零的常数 r。如果第一项是 a_1，第 k 项是

Each term in a geometric sequence is its previous term multiplied by a non-zero constant r. With an initial term a_1, the k-th term is

$$a_k = r\,a_{k-1} = a_1\,r^{k-1}.$$

等比系列是所有各项的和:

A geometric series is the sum of all elements:

$$S_N = \sum_{k=1}^{N} a_k = \sum_{k=1}^{N} a_1\,r^k$$
$$= a_1\,\frac{1 - r^{N+1}}{1 - r}.$$

92

将系列 S_N 乘以 r 然后与原系列相减，就可以得到上面的结果。

We obtain this by multiplying S_N by r then subtract it from the original series.

$$S_N = a_1 + a_1 r + a_1 r^2 + \cdots + a_1 r^N$$
$$- rS_N = a_1 r + a_1 r^2 + \cdots + a_1 r^N + a_1 r^{N-1}$$
$$(1 - r) S_N = a_1 (1 - r^{N+1}).$$

一个无穷项的等比系列，如果 $|r| < 1$，$N \to \infty$ 时 $r^{N+1} \to 0$，那么

For an infinite geometric series, if $|r| < 1$, $r^{N+1} \to 0$ as $N \to \infty$, then

$$S = S_{N \to \infty} = \frac{a_1}{1 - r}.$$

例如，

For example,

$$S = \frac{1}{2} + \frac{1}{4} + \frac{1}{8} + \frac{1}{16} + \cdots$$

$$= \frac{\dfrac{1}{2}}{1 - \dfrac{1}{2}} = 1.$$

如果 $|r| > 1$，当 $N \to \infty$，$r^{N+1} \to \infty$，$S \to \infty$。这种情形下的等比数列是发散的，不收敛。

If $|r| > 1$, $r^{N+1} \to \infty$ as $N \to \infty$, $S \to \infty$. Under this condition the geometric sequence is divergent, not convergent.

调和数列 Harmonic Sequences

$a_n = a_1 + (n-1)d$ 是一个等差数列，其倒数 $\dfrac{1}{a_n}$ 组成一个新的序列，叫做调和数列。例如

$1, 2, 3, 4, \cdots$ 是一个等差数列，

$\dfrac{1}{1}, \dfrac{1}{2}, \dfrac{1}{3}, \dfrac{1}{4}, \cdots$ 是一个调和数列。

$a_n = a_1 + (n-1)d$ is an arithmetic sequence, the reciprocals $\dfrac{1}{a_n}$ form a new sequence, called harmonic sequence. For example

$1, 2, 3, 4, \cdots$ is an arithmetic sequence,

$\dfrac{1}{1}, \dfrac{1}{2}, \dfrac{1}{3}, \dfrac{1}{4}, \cdots$ is a harmonic sequence.

斐波纳契数列Fibonacci Numbers

斐波纳契数列中的每一项是前两项的和：

$F_n = F_{n-1} + F_{n-2}$, $n > 2$ 。

如果第一第二项是 0 和 1，数列是

0, 1, 1, 2, 3, 5, 8, 13, 21, 34, 55, 89, ...

A Fibonacci sequence is a sequence in which each term is the sum of the two preceding terms:

$F_n = F_{n-1} + F_{n-2}$, $n > 2$.

Starting from 0 and 1, the sequence is

0, 1, 1, 2, 3, 5, 8, 13, 21, 34, 55, 89, ...

17　三角函数 Trigonometric Functions

锐角三角函数 Trigonometric Functions of Acute Angles

锐角 θ 的三角函数定义为直角三角形三条边(对边 a、邻边 b 和斜边 c)的比率。

For acute angles θ, the trigonometric functions are defined as ratios of a right triangle's three sides (opposite side a, adjacent side b, and hypotenuse c).

Sine
正弦函数

$$\sin \theta = \frac{a}{c}$$

Cosine
余弦函数

$$\cos \theta = \frac{b}{c}$$

Tangent
正切函数

$$\tan \theta = \frac{a}{b}$$

Cosecant
余割函数

$$csc\ \theta = \frac{c}{a}$$

Secant
正割函数

$$sec\ \theta = \frac{c}{b}$$

Cotangent
余切函数

$$\cot \theta = \frac{b}{a}$$

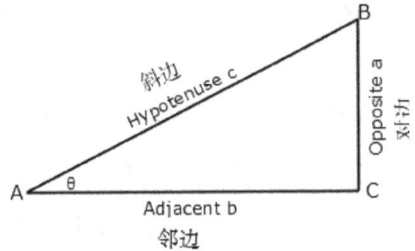

Figure 23: Right triangles. 直角三角形。

根据定义，我们得到下列关系：

From the definition, we can see the following relations:

$$\sin\theta = \frac{a}{c} = \frac{1}{csc\ \theta}\ ,\ \cos\theta = \frac{b}{c} = \frac{1}{sec\ \theta}\ ,\ \tan\theta = \frac{a}{b} = \frac{1}{\cot\theta}\ ,$$

$$\tan\theta = \frac{a}{b} = \frac{\sin\theta}{\cos\theta} = \frac{sec\ \theta}{csc\ \theta}\ ,\ \cot\theta = \frac{b}{a} = \frac{\sin\theta}{\cos\theta} = \frac{csc\ \theta}{sec\ \theta}\ .$$

因为直角三角形中的两个锐角的和是90°，所以:

Also because the two acute angles in a right triangle add up to 90°, thus:

$$\sin\theta = \frac{a}{c} = \cos\left(90° - \theta\right),\ \cos\theta = \frac{b}{c} = \sin\left(90° - \theta\right),$$

$$\tan\theta = \frac{a}{b} = \cot\left(90° - \theta\right),\ \cot\theta = \frac{b}{a} = \tan\left(90° - \theta\right),$$

$$sec\ \theta = \frac{c}{b} = csc\left(90° - \theta\right),\ csc\ \theta = \frac{c}{a} = sec\left(90° - \theta\right).$$

因为 $a^2 + b^2 = c^2$，所以

$$\left(\sin\theta\right)^2 + \left(\cos\theta\right)^2 = \frac{a^2}{c^2} + \frac{b^2}{c^2} = 1 \ 。$$

$\left(\sin\theta\right)^2$、$\left(\cos\theta\right)^2$ 常写成 $\sin^2\theta$、$\cos^2\theta$.上述公式写成

$$\sin^2\theta + \cos^2\theta = 1 \ 。$$

Because $a^2 + b^2 = c^2$, we also have

$$\left(\sin\theta\right)^2 + \left(\cos\theta\right)^2 = \frac{a^2}{c^2} + \frac{b^2}{c^2} = 1.$$

$\left(\sin\theta\right)^2$, $\left(\cos\theta\right)^2$ are written as $\sin^2\theta$, $\cos^2\theta$. The above identity is written as

$$\sin^2\theta + \cos^2\theta = 1.$$

Table 5: Values of trigonometric functions with special angles.
特殊角度的三角函数值。

	30°	45°	0°	90°
sin	1/2	$1/\sqrt{2}$	0	1
cos	$\sqrt{3}/2$	$1/\sqrt{2}$	1	0
tan	$1/\sqrt{3}$	1	0	∞
cot	$\sqrt{3}$	1	∞	0
sec	$2/\sqrt{3}$	$\sqrt{2}$	1	∞
csc	2	$\sqrt{2}$	∞	1

极坐标系 Polar Coordinates

在极坐标系中，一个点的位置由其到原点 O 的距离以及与参照方向 (OA) 的角度确定（图 24）。对于 B 点，极坐标系记为 (r, θ)，其中 r 是该点到原点 O 的距离，θ 是 ∠AOB 的角度。在极坐标系中，角度 θ 不仅仅限于锐角，它可以是任何实数值。根据定义，转一整圈的角度为 360°。

In the polar coordinate system each point is determined by its distance to the origin O, and an angle from a reference direction (OA) (Figure 24). For the point B, the polar system coordinates are denoted as (r, θ), where r is the distance between the point to the origin O, θ is the angle $\angle AOB$. In the polar system, the angle θ is not limited to acute; it can be any real value. By definition, one complete turn is 360°.

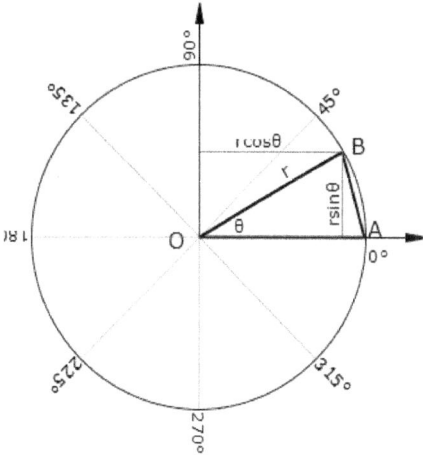

Figure 24: Polar coordinates (r, θ).
极坐标 *(r, θ)*。

如果我们将笛卡尔坐标 (x, y) 和极坐标 (r, θ) 重叠，我们会发现以下对所有实数角度 θ 都成立的关联：

If we overlap the Cartesian coordinates (x, y) and the polar coordinates (r, θ), we find for all real values of θ, the following relationships:

$$x = r \cos \theta ,$$
$$y = r \sin \theta ,$$
$$r = \sqrt{x^2 + y^2} .$$

如果 $r = 1$，我们就得到三角函数的单位圆定义：

If we set $r = 1$, we have the unit circle definition of trigonometric functions:

$$\sin \theta = y \,, \ csc \ \theta = \frac{1}{y} \,,$$

$$\cos \theta = x \,, \ sec \ \theta = \frac{1}{x} \,,$$

$$\tan \theta = \frac{y}{x} \,, \ \cot \theta = \frac{x}{y} \,,$$

这里 (x, y) 是单位圆上的一个点。这个定义将三角函数的域值扩展到所有的实数。

where (x, y) is a point on the unit circle. This definition extends the domain of trigonometric functions to all real numbers.

弧度Radian Angle Measure

弧度 (rad) 是一种角度单位。它利用圆弧的长度来测量相应的角度。 1 弧度等于其对应的弧长恰好为半径的角度（图 25）。

如果 *arc AB = r* ， *θ = 1 rad* 。

因为圆周长是2 π r ，所以一个整圈的角度是2 π rad。整圈又是360°，因此 1 rad 大约是57.296°（ 360 °/(2 π) ）。

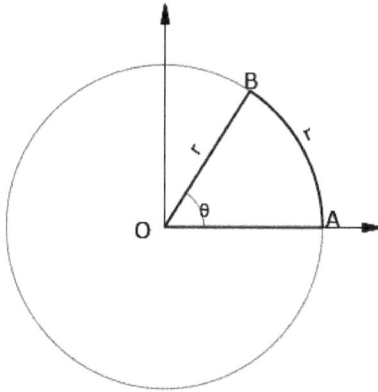

Figure 25: Definition of 1 radian
(rad) unit. 1 弧度单位的定义。

Radian (rad) is a unit of angle. It uses the length of an arc to measure the corresponding angle. 1 Radian equals to the angle where its corresponding arc length is exactly the radius (Figure 25).

θ = 1 rad when arc AB = r .

Since the circumferences of a circle is $2 \pi r$, so a complete turn is 2π rad. A complete turn is also 360°, so 1 rad is approximately 57.296° (360 °/(2 π)).

使用 rad 单位时，经常使用 π 的分数。下表（表 6）列出了一些 rad 和度之间的转换。

When using rad units, it is frequently used as fractions of π. The table below (Table 6) lists a few conversions between rad and degree angular measures.

Table 6: Conversion between rads and degrees. 弧度和度的转换。

Rad 弧度	Degree 度
2 π	360°
π	180°
π/2	90°
π/3	60°
1	57.296°
π/4	45°
π/6	30°

三角函数的周期性Periodicity of Trigonometric Functions

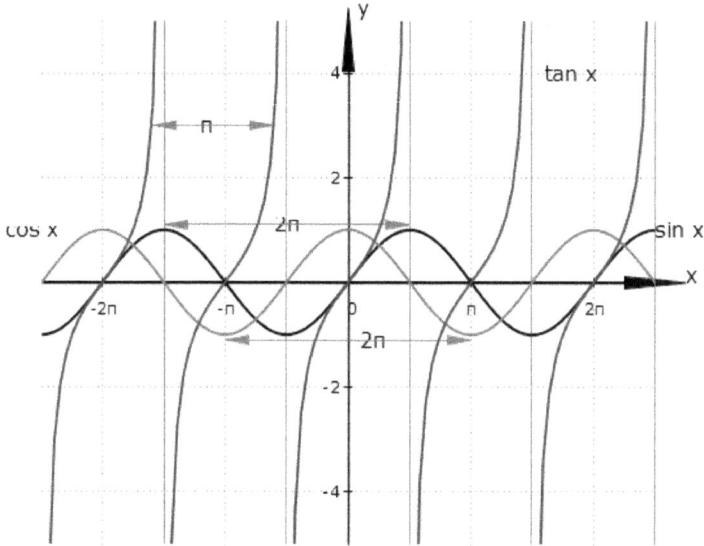

Figure 26: Trigonometric functions have periodicity of 2π or π. 三角函数的周期: 2π 或者 π。

三角函数是周期函数，它们都具有以下性质：$f(x + 2k\pi) = f(x)$，其中 x 是以弧度表示的角度，k 是整数。图 26 画出了三个三角函数的图形：$sin\ x$、$cos\ x$ 和 $tan\ x$。我们可以看出 sin 和 cos 函数的周期性为 2π，而 tan（和 cot）函数的周期性为 π。

Trigonometric functions are periodic functions, all of them have the property: $f(x + 2k\pi) = f(x)$, where x is an angle in radians and k is an integer. Figure 26 plots graphs for three trigonometric functions: *sin x*, *cos x* and *tan x*. One can see *sin* and *cos* functions have a periodicity of 2π, while *tan* (and *cot*) has a periodicity of π.

102

Table 7: Periodicity and symmetry of trigonometric functions.
三角函数的周期性和对称性。

$x + 2k\pi$	$- x$
$\sin(x + 2k\pi) = \sin x$	$\sin(-x) = -\sin x$
$\cos(x + 2k\pi) = \cos x$	$\cos(-x) = \cos x$
$\tan(x + 2k\pi) = \tan x$	$\tan(-x) = -\tan x$
$\cot(x + 2k\pi) = \cot x$	$\cot(-x) = -\cot x$
$sec(x + 2k\pi) = sec\, x$	$sec(-x) = sec\, x$
$csc(x + 2k\pi) = csc\, x$	$csc(-x) = -csc\, x$
$\pi \pm x$	$\pi/2 \pm x$
$\sin(\pi \pm x) = \mp \sin x$	$\sin(\pi/2 \pm x) = \cos x$
$\cos(\pi \pm x) = \pm \cos x$	$\cos(\pi/2 \pm x) = \mp \sin x$
$\tan(\pi \pm x) = \pm \tan x$	$\tan(\pi/2 \pm x) = \mp \cot x$
$\cot(\pi \pm x) = \pm \cot x$	$\cot(\pi/2 \pm x) = \mp \tan x$
$sec(\pi \pm x) = \pm sec\, x$	$sec(\pi/2 \pm x) = \mp csc\, x$
$csc(\pi \pm x) = \mp csc\, x$	$csc(\pi/2 \pm x) = sec\, x$

三角恒等式 Trigonometric Identities

勾股恒等式 The Pythagorean formula for sines and cosines

$$\sin^2\theta + \cos^2\theta = 1$$
$$1 + \tan^2\theta = sec^2\,\theta$$
$$1 + \cot^2\theta = csc^2\,\theta$$

和差公式 The sum and difference formulas

$$\sin(\alpha + \beta) = \sin\alpha\cos\beta + \cos\alpha\sin\beta$$
$$\sin(\alpha - \beta) = \sin\alpha\cos\beta - \cos\alpha\sin\beta$$
$$\cos(\alpha + \beta) = \cos\alpha\cos\beta - \sin\alpha\sin\beta$$
$$\cos(\alpha - \beta) = \cos\alpha\cos\beta + \sin\alpha\sin\beta$$
$$\tan(\alpha + \beta) = \frac{\tan\alpha + \tan\beta}{1 - \tan\alpha\tan\beta}$$
$$\tan(\alpha - \beta) = \frac{\tan\alpha - \tan\beta}{1 + \tan\alpha\tan\beta}$$

倍角公式 Double angle formulas

$$\sin 2\alpha = 2\sin\alpha\cos\alpha$$
$$\cos 2\alpha = \cos^2\alpha - \sin^2\alpha$$
$$\tan 2\alpha = \frac{2\tan\alpha}{1 - \tan\alpha}$$

三角形 Triangles

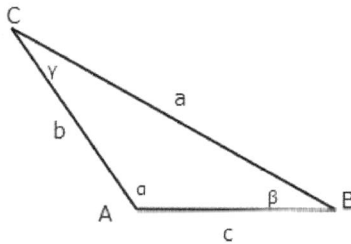

Figure 27: Sides and angles of a triangle. 三角形的边和角。

　三角形有三个边（a、b、c）和三个角（α、β、γ），它们不是相互独立的。只需其中的三到四个，包括至少一条边的长度，就足以确定三角形的整体形状和大小。以下的恒等式可用于求解未知的边长或角度。

A triangle has three sides (*a, b, c*) and three angles (α, β, γ), they are inter-dependent. Only three or four of them, including at least one side length, will be sufficient to determine the overall shape and size of the triangle. The following identities can be used to find unknown side lengths or angles.

Table 8: Laws related to a triangle. 与三角形相关的定理。

| 三角形内角和定理
The triangle sum theorem

$$\alpha + \beta + \gamma = \pi$$

正弦定理Sine Law

$$\frac{a}{\sin \alpha} = \frac{b}{\sin \beta} = \frac{c}{\sin \gamma}$$ | 余弦定理Cosine Law

$$a^2 = b^2 + c^2 - 2\,bc\,\cos \alpha$$
$$b^2 = a^2 + c^2 - 2\,ac\,\cos \beta$$
$$c^2 = b^2 + a^2 - 2\,ba\,\cos \gamma$$

正切定理Tangent Law

$$\frac{a - b}{a + b} = \frac{\tan\left(\dfrac{\alpha - \beta}{2}\right)}{\tan\left(\dfrac{\alpha + \beta}{2}\right)}$$

其他边类似 Similar for other sides |

复平面 Complex Plane

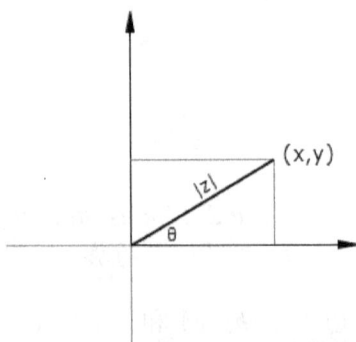

Figure 28: Complex plane. 复平面。

复数 z 可以分为实数部分 x 和虚数部分 y,

$$z = x + i\,y \text{ 。}$$

因此复数可以用笛卡尔平面（又叫复数平面）上的一点 (x, y) 来表示。转换到极坐标，我们得到

$$(r, \theta) = \left(\sqrt{x^2 + y^2}, \arctan \frac{y}{x} \right) \text{ 。}$$

复数还可以写为

$$z = x + i\,y = |z| \left(\cos\theta + i\sin\theta \right),$$

这里 $|z| = \sqrt{x^2 + y^2}$ 是 z 的绝对值。

A complex number z, can be separated in its real part x and imaginary part y as

$$z = x + i\,y.$$

It can then be represented by the point (x, y) in the Cartesian plane, also called the complex plane. Converting to polar coordinates, we have

$$(r, \theta) = \left(\sqrt{x^2 + y^2}, \arctan \frac{y}{x} \right).$$

We can rewrite the complex number as

$$z = x + i\,y = |z| \left(\cos\theta + i\sin\theta \right),$$

where $|z| = \sqrt{x^2 + y^2}$ is the absolute value of z.

z 常常写成 $z = |z|\,e^{i\theta}$ 。要严格证明

$$e^{i\theta} = \cos\theta + i\sin\theta$$

需要我们还没有学过的数学工具。下面我们验证它是正确的。

z is frequently written as $z = |z|\,e^{i\theta}$. To prove rigorously that

$$e^{i\theta} = \cos\theta + i\sin\theta$$

requires mathematical tools we have not learned. Below verify that it is true.

(1) $\theta = 0,$ $e^{i\,0} = 1$, and $\cos 0 + i \sin 0 = 1$, $e^{i\,0} = \cos 0 + i \sin 0$.

$$e^{-i\theta} = \cos \theta - i \sin \theta,$$

(2) $e^{i\theta} e^{-i\theta} = e^{i\theta - i\theta} = e^0 = 1$,

$(\cos \theta + i \sin \theta)(\cos \theta - i \sin \theta) = \cos^2 \theta + \sin^2 \theta = 1.$

$$e^{2i\theta} = \cos 2\theta + i \sin 2\theta,$$

(3) $e^{i\theta} e^{i\theta} = (\cos \theta + i \sin \theta)^2$

$$= \cos^2 \theta + 2i \cos \theta \sin \theta - \sin^2 \theta$$

$$= \cos 2\theta + i \sin 2\theta.$$

如果 $\theta = \pi$，$\cos \pi + i \sin \pi = -1$.我们得到欧拉方程，科学中最美的方程之一：

If $\theta = \pi$, $\cos \pi + i \sin \pi = -1$. We have the Euler's equation, one of the most beautiful equations in science:

$$e^{i\pi} + 1 = 0.$$

18 概率和统计 Probability and Statistics

描述统计Descriptive Statistics

Table 9: Statistics measures. 统计指标。

Name	Formula	Usage
Mean 平均值	$\overline{x} = \dfrac{1}{N} \displaystyle\sum_{k=1}^{N} x_k$	To measure and summarize the center of a data set.
Median 中位数	Middle value of an ordered data set 按大小排列的数据的中位值	测量和总结数据集的中心。
Mode 众数	Most frequent value in a data set 数据集中最常出现的数据	
Minimum 最小值	Smallest value in a data set 数据集中的最小值	To measure and summarize the spread of a data set.
Maximum 最大值	Maximum value in a data set 数据集中的最大值	
Range 极差	Maximum − Minimum 最大值−最小值	测量和总结数据集的分布。
Percentile 百分位	Order data according to their values, the k-th percentile is the value below which k percents of values fall. 对数据从小到大排序，第 k 个百分位是 $k\%$ 数据小于该值。	
Variance 方差	$\sigma^2 = \dfrac{1}{N} \displaystyle\sum_{k=1}^{N} \left(x_k - \overline{x}\right)^2$	
Standard Deviation 标准差	σ	

描述统计（叙述统计），是对数据的总结和描述。给出一组数据，我们要做的第一件事就是总结信息并描述重要特征。对于一组数字数据

$$\{x_k\}\ k=1,\cdots N\ ,$$

表 9 列出了用于总结数据集的各种度量。

Descriptive statistics, also known as narrative statistics, is the summary and description of data. Giving a collection of data, the first thing we do is to summarize the information and describe the important features. For numerical data

$$\{x_k\}\ k=1,\cdots N\ ,$$

the measures used to summarize the dataset is listed in Table 9.

例如，一支篮球队有 13 名球员。他们每场比赛的平均得分是 {15.3, 12.6, 11.4, 11, 8.9, 6.8, 6.3, 5.3, 2.8, 1.4, 1.4, 0.3, 0}。球队的得分统计表 10 中给出了总结。

For example, a basketball team has 13 players. Their per game average scores are {15.3, 12.6, 11.4, 11, 8.9, 6.8, 6.3, 5.3, 2.8, 1.4, 1.4, 0.3, 0}. Team statistics is summarized in Table 10.

Table 10: The basketball team's player score statistics. 篮球队队员得分统计。

Mean 平均值	$(15.3 + 12.6 + 11.4 + 11 \cdots + 1.4 + 0.3 + 0)/13 =$
Median 中位数	Player number N = 13 is odd, median is the center value (7th place) 6.3. If N is even, take the two center values and calculate the average. 队员数 N = 13 是奇数，中位值是中间值（第七位）。如果是偶数，取中间两数的平均值。
Mode 众数	1.4. This score appears twice, every other score only once. 这个得分出现两次，其他都只有一次。
Minimum 最小值	0
Maximum 最大值	15.3
Range 极差	15.3 − 0 = 15.3
75th Percentile 百分位	75%*13 = 9.75. The 10th highest score has 75% lower scores. 75%-Percentile is the 10th highest score which is 11. 75%的得分小于第十高的得分，75%百分位是第十高得分，得分为 11。
Variance 方差	$\sigma^2 = \dfrac{1}{N} \sum\limits_{k=1}^{N} \left(x_k - \bar{x}\right)^2 = 26.05.$
Standard Deviation 标准差 / 均方差	$\sigma = \sqrt{\sigma^2} = 5.1$

　　除了百分位数之外，四分位数也用于将数据分为四等份。第一个四分位是第 25 个百分位数，第二个四分位是第 25 到第 50

个百分位数，第三个四分位是第 50 到第 75 个百分位数，其余的是第 4 个四分位。

四分位数也可以指四分位数值。第一个四分位数值为第 25 个百分位数，第二个四分位数值为第 50 个百分位数，第三个四分位数值为第 75 个百分位数，第四个四分位数值为最大值。

In addition to percentile, quartiles are also used to divide data into four groups. The first quartile is the 25th percentile, the second quartile is the 25th percentile to the 50th percentile, and the third quartile is the 50th percentile to the 75th percentile, and the remaining is the 4th quartile.

Quartiles may also refer to quartile values. The first quartile value is the 25th percentile, the second quartile value is the 50th percentile, the third quartile value is the 75th percentile, and the 4th quartile value is the maximum.

图 示 法 探 索 数 据 Exploring Data By Graphing

　　图示经常用于直观地展示统计数据。他们可以快速探索数据并突出显示其重要特征。这里我们简单介绍一下五种常用的图示类型。

Graphs are frequently used to display data statistics visually. They can explore data quickly and highlight important features. Here we briefly introduce five commonly used graph types.

条形图Bar charts

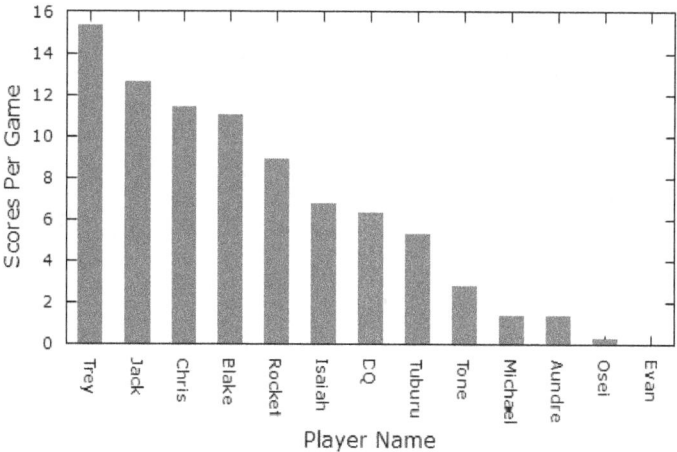

Figure 29: Basketball team player average scores per game. 篮球队队员每场比赛平均得分。

条形图用于比较不同类别的数值数据。图 29 的例子显示了一个篮球队的球员得分。

Bar charts are used to compare numerical data in different categories. For example, Figure 29 shows a basketball team's player scores.

箱线图Box plots

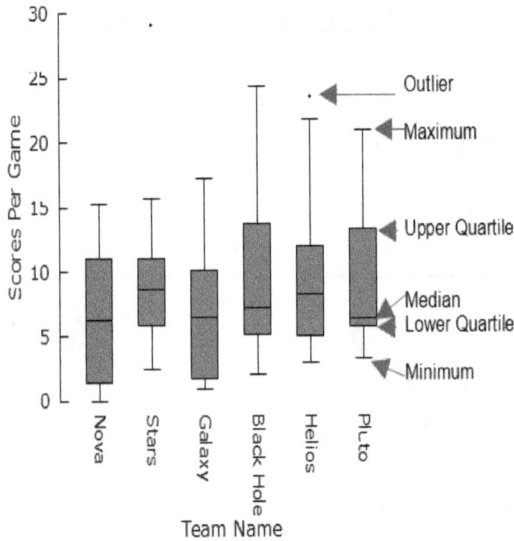

Figure 30: Team score statistics of six basketball teams. 六支篮球队的球队得分统计。

箱线图，也称为盒须图，用五个数字来总结一组数据：样本最小值、下四分位数、中位数、上四分位数和样本最大值（图 30）。与其他数据显着不同的数据点（称为异常值）显示为点。

Box plot, also called box-and-whisker plot summarizes a set of data using five numbers: the sample minimum, the lower quartile, the median, the upper quartile and the sample maximum (Figure

30). Data points that are significantly different from the rest of the data, called outliers, are displayed as dots.

直方图Histograms

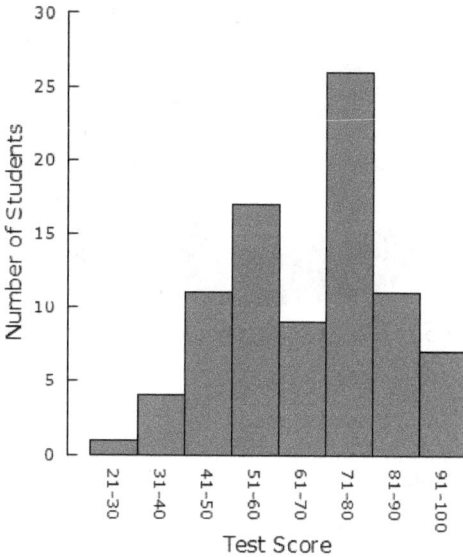

Figure 31: Statistics of students' test score.
学生的考试成绩统计。

与条形图类似，直方图使用矩形条显示数值数据。不同之处在于直方图中的类别（x轴）是数值范围（图31）。

Similar to bar chart, histograms display numerical data using rectangular bars. The difference is that categories (x-axis) in histograms are numerical ranges (Figure 31).

饼图Pie charts

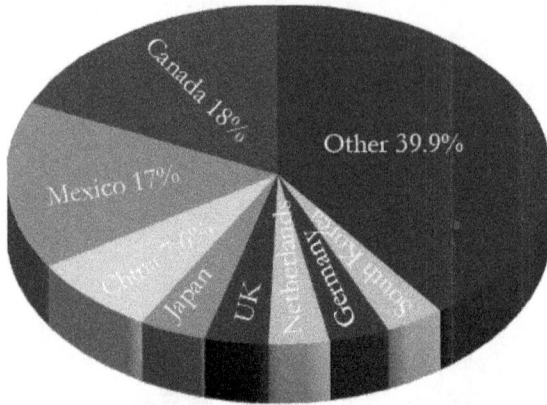

*Figure 32: US exports by country. 美国出口
各国的比例。*

饼图显示整体的分数，每个扇形的面积与其代表的分数成正
比（图 32）。例如，美国出口到墨西哥为总出口的 17%，扇形
角度为 0.17 * 360° = 61.2°。

Pie chart shows fractions of a whole, with the area of each
slice proportional to the fraction it represents (Figure 32). For
example, the US export to Mexico account 17% of total, the
angle of the slice is 0.17 * 360° = 61.2°.

散点图 Scatter Plots

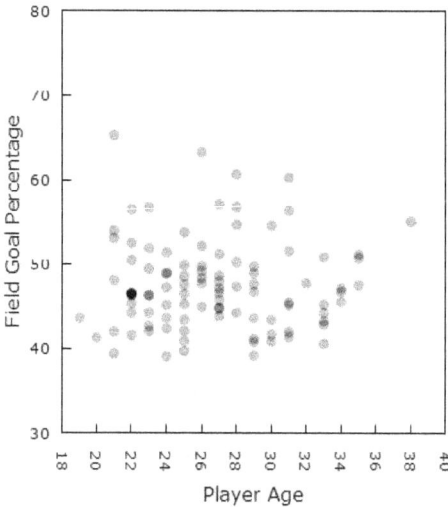

Figure 33: Basketball players' field goal percentage vs player age. 篮球队员的投中率与球员年龄。

散点图使用点表示一对数值数据（图33）。这是一种快速直观地显示一个变量受另一个变量影响程度的方法。

Scatter plot uses dots to represent a pair of numerical data (Figure 33). It is a quick way to visualize how much one variable is affected by the other.

数据采样 Data Sampling

统计学感兴趣的对象或事件的集合称为统计总体。例如，如果我们对我们班的考试成绩感兴趣，那么班级就是我们的统计

116

总体。如果我们对宇宙中中子星与恒星的比例感兴趣，那么宇宙中的所有恒星都是统计总体。显然后一种情况，我们无法试图观测宇宙中的每一颗星星来找出有多少颗恒星是中子星。宇宙中有千亿万亿颗恒星，数量巨多，我们不可能一一测量。唯一的选择是挑选一小部分恒星，并通过研究该子集来推断出可能对所有恒星都有效的结论。被选择详细研究的子集称为统计样本。

例如，大多数人用手都有偏好，有的喜欢使用右手，有的喜欢使用左手。我们想了解特定国家/地区（例如英国）的左右手使用偏好的百分比，统计总体将是所有居住在英国的人，我们实际调查的人将成为样本。调查结果如表11所示。

In statistics, the collection of objects or events we are interested in is called population. For example, if we are interested in test scores of our class, the class is our population. If we are interested in what percentage of stars are neutron stars in the universe, all stars in the universe are the population. Apparently in the latter case, we can't try to detect every star in the universe to find out how many stars are neutron star. There are hundred billion trillions of stars in the universe, too many for us to measure one-by-one. The only option is to select a small subset of stars, and by studying the small subset, to infer a conclusion that is likely valid for all stars. The subset we choose to study in detail is called a statistical sample.

For example, most people have a hand preference. They either prefer to use the right hand or the left hand. We are interested in finding out what is the percentage of hand preference in a specific country, let's say United Kingdom. The population for this inquiry will be all people living in UK. The people we actually surveyed will be the sample. The survey result is listed in Table 11.

Table 11: A sample of hand use preference in UK. 英国人惯用手样本。

Hand use 惯用手	Number of People 人数	Percentage 百分比
Right-handed 右撇子	446	88.8%
Left-handed 左撇子	47	9.4%
Equal-handed 左右平等	9	1.8%
TOTAL 总计	502	100%

分析样本得到的结果将用作对整个人口的估计，即在英国大约 88.8% 的人是右撇子，等等结论。

这项调查有三个答案，或称为结果。每个结果，右撇子、左撇子和左右平等，称为一个统计事件。所有可能的事件形成一个集合，{右撇子、左撇子、左右平等}，称为样本空间。对于样本空间中的每个事件 A，该事件的概率为

$$P(A) = \frac{A出现的次数}{样本总数} \ .$$

Results from the sample will be used as an estimate for the population, i.e., in UK approximately 88.8% people are right-handed, and so on.

For this survey, there are three answers, or called outcomes. Each outcome, right-handed, left-handed, or equal-handed, is called an event. All possible events form a set, {right-handed, left-handed, equal-handed}, called sample space. For each event A in sample space, the probability for the event is

$$P(A) = \frac{\text{number of times A occurs}}{\text{total sample size}} \ .$$

概率规则Probability Rules

规则1 Rule 1.

对任意统计事件 A, , $0 \leq P(A) \leq 1$ 。几率（机率，概率）为0表示不可能事件，几率为1表示一定会发生的事件。

For any event A, $0 \leq P(A) \leq 1$. Probability 0 means impossible event, 1 certain event.

规则2 Rule 2.

所有几率之和等于1。在上面的左右手例子中，

$P(右撇) + P(左撇) + P(左右平等) = 1$.

The sum of all the probabilities is equal to 1. In the hand example above, we have

$P(Right\text{-}handed) + P(Left\text{-}handed) + P(Even\text{-}handed) = 1$.

规则3 对立事件规则 Rule 3. Complement Rule.

一个事件 A 的对立事件（余事件，逆事件）是所有 A 不发生的事件总集，写作 A^c 或者 A'，读作 A 不发生或者非 A.

$P(A) + P(A^c) = 1$ 。

在上述手的例子中，$P(右撇) = 0.888$.

$P(非右撇) = 1 - P(右撇) = 0.112$.

Complement of an event A consists of all outcomes that are not in A, denoted as A^c or A', read as complement of A or not A.

$P(A) + P(A^c) = 1$.

In the above example, $P(Right\text{-}handed) = 0.888$.

$P(Not\ Right\text{-}handed) = 1 - P(Right\text{-}handed) = 0.112$.

规则4 加法规则 Rule 4. Addition Rule

两个互不相容（不相交）的事件 A 和 B，其中任一事件发生的概率为

$P(A \cup B) = P(A \text{ 或 } B) = P(A) + P(B)$。

如果 A 和 B 是复合事件（并非互不相容），

$P(A \text{ 或 } B) = P(A) + P(B) - P(A \text{ 和 } B)$。

$P(A \text{ 和 } B)$, 也写成 $P(A \cap B)$，是 A 和 B 两个事件都发生的几率。如果 A 和 B 互不相容，$P(A \cap B)$ 值为 0。

For two mutually exclusive (disjoint) event A and B, the probability that either one occurs is

$P(A \cup B) = P(A \text{ or } B) = P(A) + P(B)$.

If A and B are compound (not mutually exclusive),

$P(A \text{ or } B) = P(A) + P(B) - P(A \text{ and } B)$.

$P(A \text{ and } B)$, also written as $P(A \cap B)$, is the probability that event A and B both occurs. If A and B are mutually exclusive, then $P(A \cap B)$ is 0.

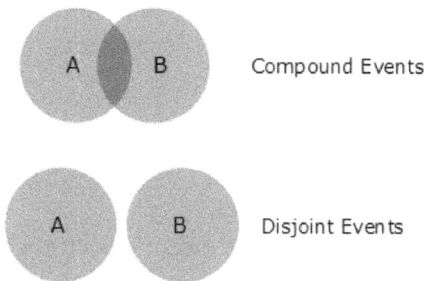

A B Compound Events

A B Disjoint Events

Figure 34: Venn diagram of compound events and disjoint events. 复合事件和不相容事件的文氏图。

规则 5 乘法规则 Rule 5. Multiplication Rule

120

$P(A \cap B) = P(A) * P(B)$ 如果事件 A 和 B 是独立事件

$P(A \cap B) = P(A) * P(B|A)$ 如果事件 A 和 B 是相关事件

当事件 A 和 B 相关（相依）时，条件概率为

$P(A|B) = P(A \cap B) / P(B)$,

$P(B|A) = P(A \cap B) / P(A)$。

$P(A|B)$是在 B 已经发生的情况下 A 发生的概率。读作 B 条件下 A 的概率。如果两个事件互不影响，那么两个事件是独立的$(P(A|B) =P(A), P(B|A) =P(B))$；如果一个事件会影响另一个事件，两个事件是相关的。

因为 $P(A \cap B) = P(A) * P(B|A) = P(B) * P(A|B)$，一个条件概率可以通过另一个条件概率计算获得

$$P(A|B) = \frac{P(B|A)\,P(A)}{P(B)}\ .$$

这个公式叫做贝叶斯定理。

$P(A \cap B) = P(A) * P(B)$ if event A and B are independent

$P(A \cap B) = P(A) * P(B|A)$ if event A and B are dependent

When event A and B are dependent, the conditional probabilities are

$P(A|B) = P(A \cap B) / P(B)$,

$P(B|A) = P(A \cap B) / P(A)$.

$P(A|B)$ is the probability of A occurs given that B has already occurred. This is read as the conditional probability of A given B. Two events are independent if one event has no effect on the other event $(P(A|B) =P(A), P(B|A) =P(B))$; dependent if one event has effect on the other event.

Since $P(A \cap B) = P(A) * P(B|A) = P(B) * P(A|B)$, one can convert one conditional probability to another

$$P(A \mid B) = \frac{P(B \mid A)\, P(A)}{P(B)},$$

which is called Bayes' theorem.

频率表或列联表可用于计算条件概率。表 12 记录了对一群人的色盲状况调查。

Frequency table or contingency table can be used to calculate conditional probabilities. Table 12 shows a survey of color blindness in a group of people.

Table 12: A sample of color blindness. 色盲的样本。

	Normal 正常	Color Blind 色盲	Total 总计	Normal 正常	Color Blind 色盲	Total 总计
Male 男	205	17	222	47.1%	3.9%	51.0%
Female 女	212	1	213	48.7%	0.2%	49.0%
Total 总计	417	18	435	95.9%	4.1%	100%

表中每个格中的频率（除以总数 435）给出了联合概率。例如，P(色盲∩男性) = 3.9%。阴影区域显示边际概率，例如，P（男性）= 51.0%。

条件概率
P（色盲|男性）= P（色盲∩男性）/ P（男性）
      = 0.039 / 0.51 = 7.6%。

In this table, each cell's frequency (divided by the total 435) gives us the joint probabilities. For example, P(color blind ∩ male) = 3.9%. The shaded area shows *marginal probabilities, for example, P(male) = 51.0%.*
The conditional probability
P(color blind | male) = P(color blind ∩ male) / P(male)
          = 0.039 / 0.51 = 7.6%.

排列组合Permutations and Combinations

假设我们抛两次硬币，得到不同结果（即 1 个正面和 1 个反面）A 的概率是多少？为了回答这个问题，让我们用字母 H 表示头部，字母 T 表示尾部，列出所有可能的结果。这里有 4 种可能：

HH HT TH TT

两个结果 HT 和 TH 满足我们的要求。所以概率是

$$P(A) = \frac{2}{4} = \frac{1}{2} \text{ 。}$$

这里我们用了下列公式

$$P(A) = \frac{\text{所有正确结果}}{\text{所有可能的结果}} \text{ 。}$$

在这个具体的例子中，每次我们抛硬币都有两种可能的结果。连抛两次，因此所有可能的结果是 2 × 2，这里允许重复。

对于所需的结果，我们的要求是两次抛掷必须结果不同。所以不允许重复时，正确结果的数量是 2 × 1，因为第一次抛掷有 2 种可能性，第二次只有 1 种可能性。

Suppose we toss a coin twice, what is the probability that we have different outcomes A, i.e., 1 head and 1 tail?

To answer the question, let's list all possible outcomes using letter H for head and letter T for tail. There are 4 possibilities:

HH HT TH TT

Two outcomes HT and TH meet our requirement. So the probability is

$$P(A) = \frac{2}{4} = \frac{1}{2} \text{ .}$$

Here we used the following formula

$$P(A) = \frac{All\ Positive\ Outcomes}{All\ Possible\ Outcomes}.$$

In this specific example, each time we toss a coin, there are two possible outcomes. We toss twice, so all possible outcomes is *2 × 2* where repetitions are allowed.

For positive outcomes, our requirements are that the two tosses must be different. When repetitions are not permitted, the number of outcomes is *2 × 1*, since the first toss has 2 possibilities and the second toss has only 1 possibility.

k-排列 *k*-Permutation

如果从 n 个物体中选 k 个，不同顺序的相同物体被认为是不同的，可以使用以下公式计算所有选择的可能性的数量

$$
^n P_k = P(k, n)
= \begin{cases} n^k & \text{如果允许重复选择同一物体} \\ \dfrac{n!}{(n-k)!} & \text{如果不允许重复（选过的不放回去）} \end{cases}
$$

$^n P_k$ 读作 n 选 k 的排列。

$$n! = n(n-1)(n-2)\cdots 2\cdot 1$$

是阶乘函数，读作 n 阶乘。*2! = 2 × 1 = 2, 3! = 3 × 2 × 1 = 6, 0! = 1.*

If *k* items are chosen from a set of *n* items, same items in different order are considered different, the number of possibilities can be calculated using the formula

$$
^n P_k = P(k, n)
= \begin{cases} n^k & \text{if repetition of an item is allowed} \\ \dfrac{n!}{(n-k)!} & \text{if repetition is prohibited (no replacement)} \end{cases}
$$

$^n P_k$ is pronounced "*n* permute *k*" or *k*-permutation of *n*.

$$n! = n(n-1)(n-2)\cdots 2\cdot 1$$

is the factorial function, reads *n* factorial. *2! = 2 × 1 = 2, 3! = 3 × 2 × 1 = 6,* and *0! = 1.*

k-组合 *k*-Combination

如果从 *n* 个物体中选 *k* 个，不同顺序的相同物体被认为是相同的，可以使用以下公式计算所有选择的可能性的数量：

$$^nC_k = C(k,n) = \binom{n}{k}$$

$$= \begin{cases} \dfrac{(n-1+k)!}{(n-1)!\,k!} & \text{如果允许重复选择同一物体} \\[2mm] \dfrac{n!}{(n-k)!\,k!} & \text{如果不允许重复（选过的物体不放回去）} \end{cases}$$

$\binom{n}{k}$ 读作 *n* 选 *k*。

If *k* items are chosen from a set of *n* items, the same items in different order are considered the same, the number of possibilities can be calculated using the formula:

$$^nC_k = C(k,n) = \binom{n}{k}$$

$$= \begin{cases} \dfrac{(n-1+k)!}{(n-1)!\,k!} & \textit{if repetition of an item is allowed} \\[2mm] \dfrac{n!}{(n-k)!\,k!} & \textit{if repetition is prohibited } - \textit{no replacement} \end{cases}$$

$\binom{n}{k}$ pronounces "*n* choose *k*".

例 1. *k*–排列，允许重复

Example 1. *k*-Permutation, repetition allowed.

6 位数密码由 0 到 9 中的六个数字组成。密码总有多少种可能性？

既然没有提及复杂性规则，因此数字重复是允许的。一共有10个数字（$n = 10$），我们需要从10个数字中选择6个（$k = 6$），总可能性为 $n^k = 10^6 = 1,000,000$。

A 6 digit pass-code is comprised of 6 digits, numbers 0 to 9. How many possibilities of pass-code are there?

Since there is no mention of complexity rule, number repetition is allowed. There are 10 digits ($n = 10$), we need to select 6 from 10 ($k = 6$), total possibilities is $n^k = 10^6 = 1,000,000$.

例2. k-排列，不允许重复

Example 2. k-Permutation, repetition not allowed.

假如6位数密码有一个复杂性规则：密码的数字不能重复。规则将不允许以下密码：111111 或者 836155。这次密码总共有多少种可能性？

同样我们有 $n = 10$，$k = 6$。第一位数我们可以挑选任何数，所以有10种可能。第二位数，除了已经选过的数，我们可以选任何数，因此有9种可能。继续下去，我们得到 $10×9×8×7×6×5 = 151,200$。如果我们使用公式，得到的答案相同：

$$\frac{n!}{(n-k)!} = \frac{10!}{(10-4)!} = 151,200 \text{。}$$

这个结果明显小于例1中的结果。您明白为什么手机密码没有复杂性规则吗？复杂性规则实际上使黑客更容易猜到密码。

Suppose for 6 digit pass-code, there is a complexity rule: any digit can't be used more than once. This rule will not allow pass-code such as 111111 or 836155. How many possibilities of pass-code are there now?

Again we have $n = 10$, $k = 6$. For the first digit, we can choose any number, so there is 10 possibilities. For the second digit, we can choose any number except the one we already chosen, so there is 9 possibilities. Continue on we have $10×9×8×7×6×5 = 151,200$. If we use the formula, we have the same answer:

$$\frac{n!}{(n-k)!} = \frac{10!}{(10-4)!} = 151,200.$$

This number is significantly smaller than the result in Example 1. Do you see the reason why cell phone pass-code does not have a complexity rule? Complexity rule actually makes it simpler for hackers to guess the pass-code.

例 3. k-组合，不允许重复

Example 3. k-Combination, repetition not allowed.

集合 $\{0,1,2,3,4,5,6,7,8,9\}$，找到其 6 位数子集的数量。

6 位数子集拥有 6 个不重复的数字，顺序无关紧要。这是一个 k-组合的问题。有 6 位数的子集数量是：

Finding the number of subsets of the set $\{0,1,2,3,4,5,6,7,8,9\}$ having 6 elements.

A set of 6 elements has 6 unique numbers in any order. Therefore it is a k-combination problem. The number of sets having 6 elements is：

$$^{10}C_6 = \frac{10!}{(10-6)!\,6!} = \frac{10!}{4!\,6!} = 210.$$

二项式定理 Binomial Theorem

多项式 $(x+y)^n$ 可以用二项式定理展开：

Polynomial $(x+y)^n$ can be expanded according to the binomial theorem:

$$(x+y)^n = \sum_{k=0}^{n} \binom{n}{k} x^k y^{n-k}$$

$$= \sum_{k=0}^{n} \frac{n!}{(n-k)!\,k!} x^k y^{n-k}.$$

展开式有 *n+1* 项，第 *k* 项的系数是 $\dfrac{n!}{(n-k)!\,k!}$ 。

The expansion has *n+1* terms, the *k*-th term has coefficient $\dfrac{n!}{(n-k)!\,k!}$.

例如 For example

$$(x+y)^3 = \frac{3!}{(3-0)!\,0!} x^0 y^{3-0} + \frac{3!}{(3-1)!\,1!} x^1 y^{3-1}$$

$$+ \frac{3!}{(3-2)!\,2!} x^2 y^{3-2} + \frac{3!}{(3-3)!\,3!} x^3 y^{3-3}$$

$$= y^3 + 3xy^2 + 3x^2 y + x^3.$$

二项式系数经常表示为帕斯卡三角形，又叫杨辉三角（图 35）。从顶部开始，使用两个相邻系数来计算下一行的系数。这是因为

The binomial coefficients are frequently showing as the Pascal triangle, also called Yang Hui triangle (Figure 35). Starting from top, two neighboring coefficients are used to calculate the coefficients in the next row. It is because

$$\binom{n}{k} = \binom{n-1}{k-1} + \binom{n-1}{k}.$$

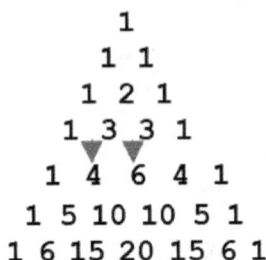

```
          1
         1 1
        1 2 1
       1 3 3 1
      1 4 6 4 1
     1 5 10 10 5 1
    1 6 15 20 15 6 1
```

Figure 35: Pascal triangle or Yang Hui triangle. 帕斯卡三角，又称杨辉三角。

概率分布Probability Distribution

样本空间是我们感兴趣的问题的所有可能结果的集合。例如，掷 6 面骰子的样本空间是集合 {1, 2, 3, 4, 5, 6}。

假设 x 是样本空间中的一个变量元素，概率分布是函数 $P(x)$，它给出了元素 x 出现的概率。在上面骰子的例子里，对于样本空间 {1, 2, 3, 4, 5, 6} 中的任何 x，$P(x)$ 是常数 1/6。

A sample space is the set that includes every possible outcomes of the problem we are interested in. For example, a sample space for rolling a 6-sided dice once is the set {1, 2, 3, 4, 5, 6}.

Let x be a variable element in the sample space, the probability distribution is the function $P(x)$, which gives the probability of occurrence of the element x. For the dice example above, $P(x)$ is a constant $1/6$ for any x in the sample space $\{1, 2, 3, 4, 5, 6\}$.

二项分布Binomial Distribution

假设一个试验只有两种可能结果：成功和失败。成功的概率为 p，试验运行 n 次，成功 x 次的概率是多少？答案是二项式分布：

$$P(x; n, p) = \binom{n}{x} p^x (1 - p)^{n - x} ,$$

这里 x 的值域为 $\{0, 1, 2, \ldots n\}$，$0 \leq p \leq 1$。

$1 - p$ 也是失败的概率 q，所以上式也可以写成

$$P(x; n, p) = \binom{n}{x} p^x q^{n - x} 。$$

所以可以这样理解：x 次成功的概率是 p^x，$n - x$ 次失败的概率是 $q^{n - x}$。由于 x 次成功可能发生在 n 次试验中的任何地方，因此有 $\binom{n}{x}$ 不同的可能。

Suppose a trial has only two outcomes, success and fail. The probability of success is p, and the trial is ran n times. What is the probability that it is success x times? The answer is the binomial distribution:

$$P(x; n, p) = \binom{n}{x} p^x (1 - p)^{n - x} ,$$

where x has possible values in $\{0, 1, 2, \ldots n\}$, $0 \leq p \leq 1$.

$1 - p$ is also the fail rate q, so the above formula can also be written as

$$P\left(x; n, p\right) = \binom{n}{x} p^x q^{n-x}.$$

It can then be understood as follows: x successes with the probability p^x, $n-x$ failures with the probability q^{n-x}. Because x successes can occur anywhere among the n trials, there are $\binom{n}{x}$ different ways.

例如，连续 5 次抛硬币，得到 4 次正面 1 次反面的概率是多少？每次正面朝上的概率为 ½，因此 $p =$ ½。失败的概率 q 也是 ½。因此我们得到

For example, if we toss a coin 5 times, what is the probability we'll get 4 head 1 tail? For coin toss, the probability of getting a head each time is ½, so $p =$ ½. The fail rate q is also ½. We thus have

$$P\left(4\right) = \binom{5}{4}\left(\frac{1}{2}\right)^4\left(\frac{1}{2}\right)^{5-4}$$
$$= \frac{5!}{4!\,1!} \times \frac{1}{32} = \frac{5}{32}.$$

5 次抛掷中至少出现 1 次正面的概率是多少？我们可以计算概率 P(1)+P(2)+P(3)+P(4)+P(5)，但还有一种更简单的方法。由于所有概率之和为 1，因此 P(1)+P(2)+P(3)+P(4)+P(5) 等于

What is the probability of at least 1 head out of 5 tosses? We can calculate the probability P(1)+P(2)+P(3)+P(4)+P(5), but there is an easier way. Since all probabilities sum to one, so P(1)+...+P(5) is exactly the same as

$$1 - P(0) = 1 - \binom{5}{0}\left(\frac{1}{2}\right)^0\left(\frac{1}{2}\right)^{5-0}$$

$$= 1 - \frac{5!}{0!\,5!} \times \frac{1}{32}$$

$$= 1 - \frac{1}{32} = \frac{31}{32}.$$

二项分布的平均值是 *np*，方差为 *np(1- p)*。

The mean of a binomial distribution is *np* and variance is *np(1-p)*.

正态分布Normal Distribution

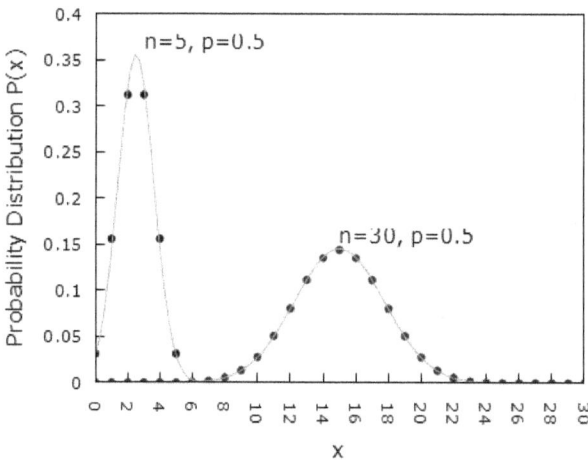

Figure 36: Binomial distribution approaches normal distribution as sample size increases. 当样本大小增加时，二项式分布趋近正态分布。

在图 36 中，我们显示了两个二项分布（$n = 5, p = 0.5$ 和 $n = 30, p = 0.5$）。二项式分布中的变量 x 是整数，因此数据用点表示。这些点似乎接近平滑且对称的曲线，n 较大的二项式数据更是与曲线完美拟合。

我们这里使用的平滑曲线称为正态分布或高斯分布，

$$P\left(x\right) = \frac{1}{\sigma\sqrt{2\pi}}\, e^{-\frac{1}{2}\left(\frac{x-\mu}{\sigma}\right)^2},$$

其中 μ 是平均值，σ 是标准差，x 是实数连续变量。正态分布也可以写成标准形式

$$z = \frac{x-\mu}{\sigma},$$

$$\phi\left(z\right) = \frac{1}{\sqrt{2\pi}}\, e^{-\frac{1}{2}z^2}。$$

In Figure 36, we show two binomial distributions ($n = 5$, $p = 0.5$ and $n = 30$, $p = 0.5$). The variable x in the binomial distribution is integer, so the data is denoted by dots. The dots seem to fit smooth and symmetric lines, with binomial data from larger n fitting perfectly.

The smooth lines we used here are called normal distribution or Gaussian distribution,

$$P\left(x\right) = \frac{1}{\sigma\sqrt{2\pi}}\, e^{-\frac{1}{2}\left(\frac{x-\mu}{\sigma}\right)^2},$$

where μ is the mean and σ is the standard deviation, and x is a continuous variable of real number. Normal distribution can also be written in a standard form as

$$z = \frac{x-\mu}{\sigma},$$

$$\phi\left(z\right) = \frac{1}{\sqrt{2\pi}}\, e^{-\frac{1}{2}z^2}.$$

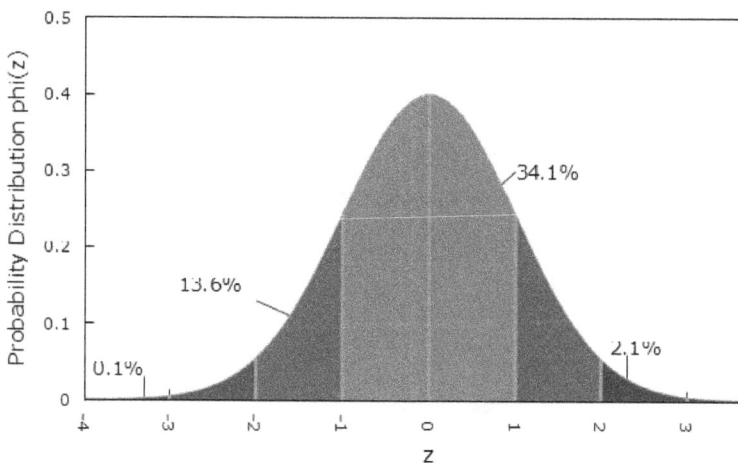

Figure 37: Standard normal distribution φ(z). 标准正态分布 *φ(z)*。

图 37 显示的是正态分布的标准形式 **φ** *(z)*。注意

$x = \mu + \sigma z$,

z = ± *1* 对应于 *x = μ ± σ*。曲线下在 *z* = ± *1* 之间的面积是 68.1%，*z* = ± *2* 之间的面积是 95%（准确地说 95.45%，对应于 95%的是 $|z| = 1.96$）。不严格地说，2-sigma（*z* = ± *2*）常被叫做 95%置信区间。

Figure 37 shows the standard normal distribution **φ** *(z)*. Note

$x = \mu + \sigma z$,

z = ± *1* corresponds to *x = μ ± σ*. The area under the curve between *z* = ± *1* is 68.1%, between *z* = ± *2* is 95% (95.45% to be exact, and $|z| = 1.96$ for 95%). Approximately, 2-sigma (*z* = ± *2*) is frequently called 95% confidence interval.

给定任意 *x*，*μ* 和 *σ*，我们可以先计算 *z* 值，然后用 *z* 值表（标准正态分布表）或者计算器找到 **φ** *(z)*。注意 **φ** *(z)*的最大值

134

是$1/\sqrt{2\pi}$，但 $P(x)$ 的最大值是$1/\left(\sqrt{2\pi}\sigma\right)$，我们需要将 $\phi(z)$ 值除以 σ 以得到 $P(x)$：

$$P(x) = \frac{\phi(z)}{\sigma}, \; z = \frac{x-\mu}{\sigma} \; 。$$

For any given value of x, μ and σ, we can first calculate the z-value, and then find $\phi(z)$ by using the z-score table (standard normal distribution table) or a calculator. Note the maximum of $\phi(z)$ is $1/\sqrt{2\pi}$, and the maximum of $P(x)$ is $1/\left(\sqrt{2\pi}\sigma\right)$, to get $P(x)$ we need to divide $\phi(z)$ by σ:

$$P(x) = \frac{\phi(z)}{\sigma}, \; z = \frac{x-\mu}{\sigma} \; .$$

假设检验Hypothesis Testing

总体与样本 Population and sample

从统计学的角度来看，概率分布是统计总体的理论模型。例如，正态分布 $P(x, \mu, \sigma)$ 给出了 x 出现在总体平均值为 μ 且总体标准差为 σ 的总体中的概率。x 是变量，μ, σ 是参数。

在大多数情况下，参数是未知的。我们的目标实际上是建立一个模型并确定这个模型是否准确。在本节中，我们讨论如何测试一个模型是否准确。

From a statistical perspective, a probability distribution is a theoretical model of a population. For an example, a normal distribution $P(x, \mu, \sigma)$ gives us the probability x occurs in a population with population mean μ and population standard deviation σ. x is a variable, μ, σ are parameters.

In most cases, the parameters are unknown. Our goal is actually to build a model and to determine whether the model is accurate. In this section, we discuss how to test whether a model is valid.

为了测试模型，我们采集样本并将样本与模型进行比较。样本是总体的一个子集。在比较之前，我们需要确保样本能够代表总体。不谈其他因素，目前我们将讨论样本大小 n 的影响。

To test a model, we take a sample and compare the sample against the model. A sample is a subset of the population. Before we proceed, we need to make sure a sample is representative of the population. There are other factors, for the moment we will discuss the effects of sample size n.

有很多方法可以从总体中获取大小为 n 的样本 X。对于每个样本，都会有一个样本均值 $\bar{x}$。描述样本的数字，例如样本均值、样本方差，称为样本统计量。考虑到取样的方法有很多种，样本均值 $\bar{x}$ 是一个随机变量。样本均值的期望均值是：

There are many ways to take a sample X of size n from the population. For each sample, there will be a sample mean $\bar{x}$. Numbers describing a sample, such as sample mean, sample variance, are called sample statistic. Considering there are many ways to take a sample, sample mean $\bar{x}$ is a random variable. The expect mean of the sample mean is:

$$E\left(\bar{x}\right) = E\left(\frac{x_1 + x_2 + \cdots + x_n}{n}\right)$$
$$= \frac{E\left(x_1\right) + E\left(x_2\right) + \cdots + E\left(x_n\right)}{n}$$
$$= \frac{\mu + \mu + \cdots + \mu}{n} = \mu .$$

样本均值的期望方差为：

The expect variance of the sample mean is:

$$Var\left(\bar{x}\right) = Var\left(\frac{x_1 + x_2 + \cdots + x_n}{n}\right)$$
$$= Var\left(\frac{1}{n}\,x_1\right) + Var\left(\frac{1}{n}\,x_2\right) + \cdots + Var\left(\frac{1}{n}\,x_n\right)$$
$$= \frac{1}{n^2}\left(Var\left(x_1\right) + Var\left(x_2\right) + \cdots + Var\left(x_n\right)\right)$$
$$= \frac{1}{n^2}\,n\,\sigma^2 = \frac{\sigma^2}{n} .$$

样本均值的均方差为：

The standard derivation of the sample mean is:

$$Std\left(\bar{x}\right)=\frac{\sigma}{\sqrt{n}}.$$

我们预计大样本将更好地代表总体。正如预期的那样，均方差随着样本量的增加而减小。

We expect large sample size will be better to represent the population. The standard derivation decreases as sample size increases as expected.

中心极限定理The central limit theorem

只要样本量足够大，样本均值的抽样分布就接近正态分布。

The sampling distribution of the sample mean $\bar{x}$ approaches the normal distribution as long as the sample size is large enough.

零假设备择假设 Null hypothesis and alternative hypothesis

样本可用于估计和建立总体模型。它还可以用来测试模型假设是否可能成立。

假设检验是一种对假设（即总体模型）做出决策的统计方法。标准方法是首先建立两个相互矛盾的假设，称为原假设（零假设、虚无假设）H_0和备择假设（对立假设、备选假设）H_a。

H_0，零假设：样本和总体之间没有显着差异，任何观察到的差异都是由于抽样误差造成的。

H_a，备择假设：样本和总体之间存在显著差异。

我们假设原假设为真，并使用样本信息来决定是否有足够的证据来拒绝原假设。

Sample can be used to estimate and build a model of the population. It can also be used to test whether a model hypothesis is likely true.

Hypothesis testing is a statistical method for making a decision about a hypothesis, namely a model of the population. The standard method is setting up two conflicting hypotheses called the null hypothesis H_0 and the alternative hypothesis H_a.

H_0, the null hypothesis: a claim there is no significant difference between sample and population, any observed difference is due to sampling errors.

H_a, the alternative hypothesis: there is significant difference between sample and population.

We presume the null hypothesis is true, and use sample information to decide if there is sufficient evidence to reject the null hypothesis.

给定H_0和H_a，有四种可能结果（表13）。

Given H_0 and H_a, there are four possible outcomes (Table 13).

Table 13: Null hypothesis and possible decision outcomes. 元假设和可能的决定结果。

H_0 Decision 决定	H_0 hypothesis is actually H_0假设事实上是	
	True 正确	False 错误
Don't reject 不拒绝	Correct decision 决定正确 True negative 1-α 真阴性 1-α	Type II error 第二类错误 False negative β 假阴性 β
Reject 拒绝	Type I error 第一类错误 False positive α 假阳性 α	Correct decision 决定正确 True positive 1-β 正阳性 1-β

在设计如何收集样本数据之前，我们不仅需要考虑第一类误差α，还需要考虑第二类误差β（统计功效1-β）。我们需要仔细考虑H_0、H_a、α和β，以便利用有限的资源提出最佳设计。

选定假设和误差极限之后，下一步就是计算 p 值。p 值是假设零假设正确，测试数据完全因为随机因素获得的概率。如果 p 值小于 α，那么测试数据不太可能是偶然机遇，我们应该拒绝零假设。α 通常设置为 0.05。它也称为显著性水平。$1 - \alpha$ 称为测试的特异性。

Before we design how to collect sample data, we need to consider not only type I error α, but also type II error β (the statistical power $1 - \beta$). We would need to carefully consider H_0, H_a, α and β together to come up with the best design using limited resources.

Once the hypotheses and error limits are chosen, the next step is to calculate the p-value. The p-value is the probability, assuming the null hypothesis is correct, of obtaining the test data by random chance. If p-value is smaller that α, then the test data is not likely due to chance, we should reject the null hypothesis. α is commonly set at 0.05. It is also called significance level. $1 - \alpha$ is called specificity of the test.

单均值的假设检验 Hypothesis testing of single mean

一家公司声称，其减重饮食可以平均减轻体重 15 磅 (μ = 15)，标准差为 20 (σ = 20)。一名学生对此说法深表怀疑，决定设法验证。他这样设置原假设和备择假设：

H_0: $\mu = 15$

H_a: $\mu < 15$

他还选择了最常用的显著性水平 0.05 (α = 0.05)。假设总体为正态分布，将使用单尾（左尾）z 检验。他找到 7 个人使用了该公司的减重饮食，平均体重减轻了 2 磅。

z 值是:
$$z = \frac{\bar{\mu} - \mu}{\frac{\sigma}{\sqrt{n}}} = \frac{2 - 15}{\frac{20}{\sqrt{7}}} = -1.59$$ 。

–1.59 大于 –1.645，后者是 5% 显著性水平时的 $z_{1-\alpha}$ 值（图 38）。p 值是 0.056，大于 $\alpha = 0.05$。因此，学生得出的结论是，零假设不能被拒绝，公司的说法有可能是正确的。

A company claims that its diets lead to on average 15 lbs of weight loss ($\mu = 15$), with a standard deviation of 20 ($\sigma = 20$). A student is suspicious of the claim, decide to test it. He setup the null hypothesis and alternative hypothesis the following way:

H_0: $\mu = 15$ H_a: $\mu < 15$

He also selects a significant level of 0.05 ($\alpha = 0.05$), which is the most commonly used. Also assumes the population is normal, one-tailed (left tailed) z-test will be used.

He found 7 people who used the company's diets and the average weight loss is 2 lbs.

The z-value is: $z = \dfrac{\bar{\mu} - \mu}{\dfrac{\sigma}{\sqrt{n}}} = \dfrac{2 - 15}{\dfrac{20}{\sqrt{7}}} = -1.59$,

which is larger than -1.645, $z_{1-\alpha}$ at 5% significance level (Figure 38). The p-value is 0.056, bigger than $\alpha = 0.05$. So the student concluded that the null hypothesis is not rejected, the company's claim may be true.

仔细检查自己的分析后，他发现了一个错误。应用 z 检验有适用条件：（1）样本均值服从正态分布，（2）总体方差已知，（3）样本量大。

学生的样本量太小。于是他找到了另一组数据，其中 35 名使用该公司减重饮食的用户报告平均体重减轻了 5 磅。再次计算 z 值为

$$z = \dfrac{\bar{\mu} - \mu}{\dfrac{\sigma}{\sqrt{n}}} = \dfrac{5 - 15}{\dfrac{20}{\sqrt{35}}} = -2.96 \text{ 。}$$

–2.96 小于 –1.645 ($z_{1-\alpha}$)（图38）。 p 值是 0.002 < 0.05。因此零假设必须被拒绝，该公司减重 15 磅的声称没有得到 35 名用户样本的支持。

After double-check his test he found a mistake. There are conditions for applying z-test: (1) The sample mean is normally distributed, (2) population variance is known, (3) the sample size is large.

The student's sample size is too small. He found another dataset where 35 users of the company's diets reported a mean weight loss of 5 lbs. Again the z-value is

$$z = \frac{\bar{\mu} - \mu}{\frac{\sigma}{\sqrt{n}}} = \frac{5 - 15}{\frac{20}{\sqrt{35}}} = -2.96 \, ,$$

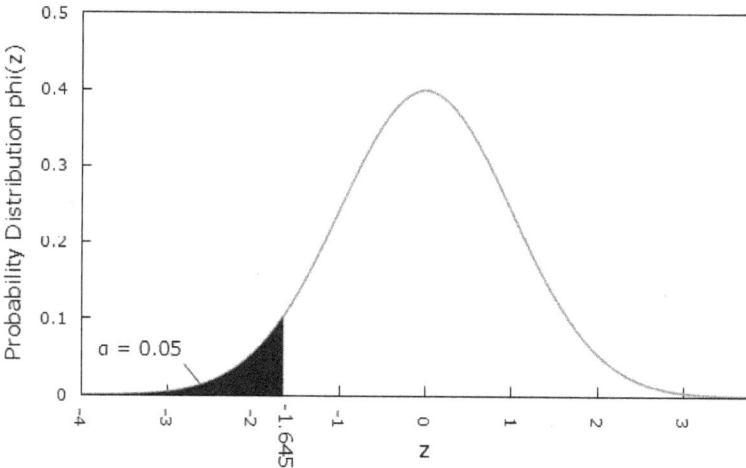

Figure 38: Standard normal distribution and one-tailed test (left-tailed). 标准正态分布和单侧检验（左尾）。

which is smaller than -1.645 ($z_{1-\alpha}$) (Figure 38). The p-value is 0.002 < 0.05. The null hypothesis is now rejected, the company's claim of 15 lb weight loss is not supported by the sample with 35 users.

19 矩阵 Matrices

矩阵运算 Matrix Operations

矩阵是将数据按行和列排列成矩形。一个 $m \times n$ 的矩阵（m 乘 n 的矩阵）写成

Matrix is a rectangular arrangement of data into rows and columns. An $m \times n$ matrix (m-by-n matrix) is written as

$$A = \begin{bmatrix} a_{11} & a_{12} & \cdots & a_{1n} \\ a_{21} & a_{22} & \cdots & a_{2n} \\ \vdots & \vdots & \ddots & \vdots \\ a_{m1} & a_{m2} & \cdots & a_{mn} \end{bmatrix}.$$

矩阵 **A** 缩写为

$$[a_{ij}] , (a_{ij}) , (a_{i,j}) \text{ 或者 } (a_{ij})_{1 \leqslant i \leqslant m, 1 \leqslant j \leqslant n} \text{ 。}$$

例如 $\begin{bmatrix} 2 & 12 & -1 \\ 35 & -6 & 11 \end{bmatrix}$ 是一个 2×3 矩阵。一行或者一列的矩阵，

类似 $\begin{bmatrix} 5 \\ 0 \\ 12 \end{bmatrix}, (0\ 0\ 0)$，有时会叫做矢量（向量）。

Abbreviated form of **A** is $[a_{ij}] , (a_{ij}) , (a_{i,j})$ or $(a_{ij})_{1 \leqslant i \leqslant m, 1 \leqslant j \leqslant n}$.

For instance $\begin{bmatrix} 2 & 12 & -1 \\ 35 & -6 & 11 \end{bmatrix}$ is a 2×3 matrix. One row or one column matrix, such as $\begin{bmatrix} 5 \\ 0 \\ 12 \end{bmatrix}, (0\ 0\ 0)$, are sometimes called vectors.

143

加法和减法 Addition and Subtraction

两个相同大小的 $m \times n$ 矩阵，在同样位置上的项相加或者相减：

For two matrices of the same size $m \times n$, add/subtract each element at the same position:

$$\left(A \pm B\right)_{i,j} = A_{i,j} \pm B_{i,j}$$

标量乘法 Scalar multiplication

标量乘法是实数常数 (c) 与矩阵 **A** 的乘积。每个元素与常数相乘，得到一个相同大小的矩阵。

Scalar multiplication is the product of a real constant (c) and a matrix **A**. Each element is multiplied by the constant, the resulting matrix is of the same size.

$$\left(c\, A\right)_{i,j} = c \cdot A_{i,j}.$$

转置 Transposition

一个 $m \times n$ 矩阵的转置是将其行和列互换，产生一个新的 n $\times$ m 矩阵，用 A^T 表示。

Transpose of an $m \times n$ matrix swaps row and column indices, producing the n $\times$ m matrix, denoted A^T.

$$\left(A^T\right)_{i,j} = A_{j,i}.$$

点积 Dot product

点积常用来衡量两个向量之间的相对方向。这里我们只考虑两个长度相等（m）的一列矩阵（**a** 和 **b**），点积为

Dot product is frequently used as a measure of relative direction of two vectors. Here we consider only two one column matrix (**a** and **b**) of equal length (m), the dot product is

$$a \cdot b = \sum_{i=1}^{m} a_i\, b_i = a^T\, b.$$

乘法 Multiplication

一个 $m \times n$ 矩阵和一个 $n \times p$ 矩阵相乘的结果是一个 $m \times p$ 矩阵。积的 ij 元素是第一个矩阵的第 i 行和第二个矩阵的第 j 列的点积。

Multiplication of an $m \times n$ matrix and an $n \times p$ matrix results in an m-by-p matrix. The product's ij element is the dot product of the i-th row of the first matrix and j-th column of the second matrix.

$$[AB]_{i,j} = a_{i,1} b_{1,j} + a_{i,2} b_{2,j} + \cdots + a_{i,n} b_{n,j}$$
$$= \sum_{k=1}^{n} a_{i,k} b_{k,j} = a_{i,} \cdot b_{,j}.$$

$$\begin{bmatrix} 2 & 1 & 0 \\ 1 & 0 & 1 \end{bmatrix} \begin{bmatrix} 1 & 2 & 1 \\ 0 & 1 & 2 \\ 1 & 2 & 2 \end{bmatrix} = \begin{bmatrix} 2 & 5 & 4 \\ 2 & 4 & 3 \end{bmatrix}$$

Figure 39: Multiplication of two matrices.
两个矩阵的乘法。

如果 $p \neq m$, **BA** 不可计算。就算 $p = m$, 一般 **AB** 和 **BA** 也不相等:

AB ≠ **BA**.

矩阵乘法遵守结合律

(**AB**)**C** = **A**(**BC**),

以及(左或者右)分配律

(**A** + **B**)**C** = **AC** + **BC**, **C**(**A** + **B**) = **CA** + **CB**.

If $p \neq m$, **BA** does not exist. Even if $p = m$, in general **AB** and **BA** are not equal:

145

$AB \neq BA.$

Matrix multiplication satisfies the associative rule

$(AB)C = A(BC),$

and the (right or left) distributive rule

$(A + B)C = AC + BC, \ C(A + B) = CA + CB.$

单位矩阵和逆矩阵Identity matrix and inverse matrix

单位矩阵是一个正方矩阵，对角线上的元素全是1，其他元素全是零。

Identity matrix is a square matrix. All its diagonal elements are 1 and all other elements are zero.

$$I_n = \begin{bmatrix} 1 & 0 & \cdots & 0 \\ 0 & 1 & \cdots & 0 \\ \vdots & \vdots & \ddots & \vdots \\ 0 & 0 & \cdots & 1 \end{bmatrix}_{n \times n}.$$

对任意一个 $m \times n$ 矩阵 A，

For any $m \times n$ matrix **A**,

$$I_m A = A I_n = A.$$

一个正方矩阵，如果存在一个正方矩阵 **B** 使得

$AB = BA = I,$

那么 **B** 是唯一的，叫做 **A** 的逆矩阵，写作 A^{-1}。**A** 称为是可逆的。

For a square matrix **A**, if there exists a square matrix **B** such that

$AB = BA = I,$

then **B** is unique and it is called inverse matrix of **A**, denoted as A^{-1}. **A** is called invertible.

多元线性方程组 Linear equations of multiple variables

多元线性方程组可以写成:

A system of linear equations can be written as:

$$a_{11} x_1 + a_{12} x_2 + \cdots + a_{1n} x_n = b_1$$
$$\vdots$$
$$a_{m1} x_1 + a_{m2} x_2 + \cdots + a_{mn} x_n = b_m$$

$$\begin{bmatrix} a_{11} & a_{12} & \cdots & a_{1n} \\ a_{21} & a_{22} & \cdots & a_{2n} \\ \vdots & \vdots & \ddots & \vdots \\ a_{m1} & a_{m2} & \cdots & a_{mn} \end{bmatrix} \begin{bmatrix} x_1 \\ x_2 \\ \vdots \\ x_n \end{bmatrix} = \begin{bmatrix} b_1 \\ b_2 \\ \vdots \\ b_m \end{bmatrix}$$

$$A\,x = b$$

如果 $m = n$, 而且各方程是相互独立的, 那么方程组的解是

If $m = n$, and the equations are independent, then the solution is

$$x = A^{-1} b \,.$$

行列式和逆矩阵 Determinant and Inverse

一个矩阵的逆矩阵可以用下列公式求得：

The inverse matrix can be found using the following formula:

$$A^{-1} = \frac{adj\,(A)}{det\,(A)},$$

这里 *adj(A)* 是伴随矩阵，*det(A)* 是矩阵 **A** 的行列式。只有当且仅当 *det(A)* ≠ *0* 时，逆矩阵才存在。

对于一个 2 × 2 的矩阵

where *adj(A)* is the adjoint of a matrix and *det(A)* is the determinant of a matrix **A**. The inverse exists if and only if *det(A)* ≠ *0*.

For a 2-by-2 matrix

$$A = \begin{bmatrix} a & b \\ c & d \end{bmatrix},$$

$$det\,(A) = \begin{vmatrix} a & b \\ c & d \end{vmatrix} = ad - bc,$$

$$adj\,(A) = \begin{bmatrix} d & -b \\ -c & a \end{bmatrix},$$

$$A^{-1} = \frac{1}{ad - bc} \begin{bmatrix} d & -b \\ -c & a \end{bmatrix}.$$

坐标变换 Coordinates transformation

将一个 xy 笛卡尔坐标系，逆时针旋转 θ 角度，我们得到一个新的 $x'y'$ 坐标系。两者与极坐标 (r, α) 的关系如下：

Given a xy-Cartesian coordinate system, by rotating counter clockwise an angle θ, we have a new $x'y'$-coordinates. Both are related to the polar coordinates (r, α) as follows:

$$x = r \cos \alpha , \ y = r \sin \alpha ;$$
$$x' = r \cos (\alpha - \theta) , \ y' = r \sin (\alpha - \theta) .$$
$$x' = r \cos \alpha \cos \theta + r \sin \alpha \sin \theta ,$$
$$y' = r \sin \alpha \cos \theta - r \cos \alpha \sin \theta ,$$
$$x' = x \cos \theta + y \sin \theta ,$$
$$y' = - x \sin \theta + y \cos \theta .$$

矩阵形式为　　　　　　　　In matrix form

$$\begin{bmatrix} x' \\ y' \end{bmatrix} = \begin{bmatrix} \cos \theta & \sin \theta \\ - \sin \theta & \cos \theta \end{bmatrix} \begin{bmatrix} x \\ y \end{bmatrix} .$$

反向为　　　　　　　　　The reverse is

$$\begin{bmatrix} x \\ y \end{bmatrix} = \begin{bmatrix} \cos \theta & \sin \theta \\ - \sin \theta & \cos \theta \end{bmatrix}^{-1} \begin{bmatrix} x' \\ y' \end{bmatrix} = \begin{bmatrix} \cos \theta & - \sin \theta \\ \sin \theta & \cos \theta \end{bmatrix} \begin{bmatrix} x' \\ y' \end{bmatrix} .$$

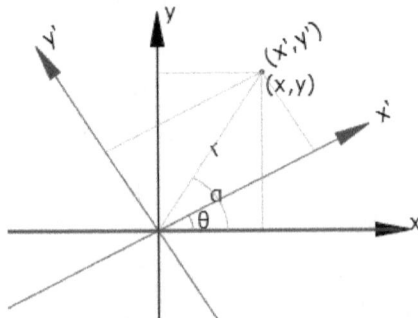

Figure 40: Rotation of reference coordinates. 参照系的旋转。

坐标变换的另一个例子是爱因斯坦的狭义相对论。狭义相对论将时间和空间联系在一起，形成了一个四维空间（闵可夫斯基空间）。这种四维空间也称为时空，它可以用四维矢量表示，

Another example of coordinates transformation is Einstein's special relativity. Special relativity links time and space together, forming the four dimensional space (Minkowski space). Also called space-time, it can be represented by a four-vector,

$$\begin{bmatrix} ct \\ x \\ y \\ z \end{bmatrix}.$$

c是光速，t是时间。我们不讨论细节，而是重点介绍如何使用矩阵来解决一个物理问题。

c is speed of light, t is time. Without going to details, we highlight how matrix is used to solve a Physics problem.

相对论原理指出，物理定律在所有参考系中都具有相同的形式，无论它们是否相对于彼此移动。根据相对论原理，以速度$\vec{v}$运动的参考系有如下变换（洛伦兹变换）

The relativity principle states laws of physics are the same form in all frames of reference, whether they are moving relative to each other. Based on the relativity principle, a reference frame moving at speed $\vec{v}$ has the following transformation (Lorentz transformation)

$$\begin{bmatrix} ct' \\ x' \\ y' \\ z' \end{bmatrix} = \begin{bmatrix} \gamma & -\beta\gamma & 0 & 0 \\ -\beta\gamma & \gamma & 0 & 0 \\ 0 & 0 & 1 & 0 \\ 0 & 0 & 0 & 1 \end{bmatrix} \begin{bmatrix} ct \\ x \\ y \\ z \end{bmatrix},$$

$$\beta = v/c, \gamma = \frac{1}{\sqrt{1-\beta^2}}.$$

150

一个运动物体的四维速度是

The four-velocity of a moving object is

$$\vec{u} = \gamma \begin{bmatrix} c \\ v_x \\ v_y \\ v_z \end{bmatrix}.$$

一个运动物体的四维动量是

The four-momentum of a moving object is

$$\vec{p} = \begin{bmatrix} \dfrac{E}{c} \\ p_x \\ p_y \\ p_z \end{bmatrix}.$$

四维动量第一项中的 E 是能量。动量与速度有如下关系：$\vec{p} = m\,\vec{v}$，这里 m 是物体的质量。四维动量与四维速度也应该有同样的关系。比较四维动量与四维速度的第一项，我们得到

E in the first term of four-momentum is energy. Momentum is related to velocity as $\vec{p} = m\,\vec{v}$, here m is the mass of the object. This should also be true for four-momentum and four-velocity. Compare the first term in four-momentum and four-velocity, we have

$$\frac{E}{c} = m\,\gamma\,c\,, \; E = \gamma\,m\,c^2 = \frac{mc^2}{\sqrt{1 - v^2/c^2}}\,.$$

如果速度 $v = 0$，我们将得到那个爱因斯坦著名的定理：

If speed $v = 0$, we have Einstein's famed formula:

$$E = mc^2\,.$$

20 词汇表 Glossary

数学符号 Math Symbols

Symbol	English	中文	Symbol	English	中文
+	plus, add	加	=	equal	等于
−	subtract, minus	减	≠	not equal	不等于
* or × or ·	multiply, time	乘	≈	approximate	约等于
exp	exponent	乘方	∝	proportional	正比
/ or ÷	divide	除	<	less than	小于
** or ^	exponent	乘方	>	greater than	大于
% or mod	modulo	取模	≤ or <=	not greater than	不大于，小于等于
$\prod\square$	product	连乘	≥ or >=	not less than	不小于，大于等于
$\sum\square$	sum up	总和	∈	is a member of	属于
%	percent	百分比	∩	intersection	交集
!	factorial	阶乘	∪	union	并集
| |	absolute value	绝对值	( )	Parentheses	括号，小括号
$\sqrt{\square}$	radical	根号	[ ]	Square brackets	中括号，方括号
π	Pi	派	{ }	Curly brackets	大括号，花括号

词汇 Vocabulary

字母顺序 Alphabetical Order

A

1D, one dimension	一维
2D, two dimension	二维
3D, three dimension	三维
Abacus	算盘
Abbreviation	缩写
Absolute value	绝对值
Acceleration	加速
Accept	接受
Accuracy	精度
Acute angle	锐角
Acute triangle	锐角三角形
Add; Plus	加
Addend	加数
Addition	加法
Additive inverse	加法逆元
Additive inverse	相反数
Adjacent	相邻
Adjoint matrix	伴随矩阵
Algebra	代数
Algorithm; method	算法
Alternative hypothesis	备择假设
Ambiguity	歧义
Amount; Quantity	量
Analog	模拟
Analyze	分析
Angle	角
Annual	每年
Answer	答案, 回答
Ante meridian (a.m.)	上午
Apex; vertex; vertices	顶点
Application	应用
Apply	运用
Approach	接近
Arabic	阿拉伯
Arc	圆弧
Area	面积
Argument	自变量
Argument	论据
Arithmetic	算术
Arithmetic progression	等差数列
Arithmetic sequence	等差数列
Arrange	排列
Array	数组
Ascending	上升
Ascending	渐升
Associative property	结合特性; 结合律
Assume	假如
Asymptote	渐近线
Attribute	属性; 归因于
Average; Mean	平均值
Axis (axes)	轴

B

Bar	杠
Bar chart	条形图
Bar graph	条形图
Base	底
Base	基数
Bayes' theorem	贝叶斯定理
Bi-	两个
Billion	十亿
Binary	二进制
Binomial	二项式

Binomial distribution	二项分布	Clockwise	顺时针
Binomial theorem	二项式定理	Closed	封闭
Bisect, half	二等分	Closed circle	实心圆
Border	边界	Codomain	对应域
Borrow	借	Coefficient	系数
Borrow; regroup	退位	Collection	收集
Bottom	底部	Color blind	色盲
Box plot	箱线图	Combination	组合
Box-and-whisker plot	盒须图	Combine	结合
		Common	共同
Brahmi	婆罗米	Common denominator	公分母
Branch	分支	Common factor	公约数；公因子
		Common multiple	公倍数
C		Commutative property	交换律
Calculate, computation	计算	Compare; Comparison	比较
Calculator	计算器	Compass	圆规
Calculus	微积分	Complement event	对立事件
Calendar	日历	Complement event	余事件
Capacity	容量	Complementary angles	余角
Capital letter	大写字母	Complete the square	配方法
Carry	进位		
Cartesian	笛卡尔	Complex number	复数
Cartesian system	笛卡尔系统	Complex plane	复平面
Category	类别	Complexity	复杂性
Celsius	摄氏度	Composite number	合数；合成数
Cent	分	Composition function	复函数
Centimeter (cm)	公分		
Centimeter (cm)	厘米	Compound event	复合事件
Central limit theorem	中心极限定理	Compound interest	复利
		Computer	计算机
Chart	图表	Concentration	浓度
Check	检查	Concentric	同心
Chip	芯片	Conclusion	结论
Choice	选择	Conditional probability	条件概率
Choose	选取		
Circle	圆	Cone	圆锥
Circle graph	圆饼图	Confidence interval	置信区间
Circular	圆形	Congruent	全等
Circumference	周长		
Circumference	圆周长		

Congruent triangles	全等三角形	Decay	衰减
Conjecture	猜想	Decimal	十进制
Conjugate	共轭	Decimal	小数
Connect	连接	Decimal point	小数点
Consecutive	连续的	Decimeter	分米
Consecutive	相邻的	Decompose	分解
Constant	常数	Decrease	减少
Construct	构造	Decreasing	递减
Contingency table	列联表	Definition	定义
Continuous	连续	Degree	度，指数
Contradiction	矛盾	Denominator	分母
Contrast	对比	Dependent events	相关事件
Convergent	收敛	Dependent variable	因变量
Convert	转换	Depth	深度
Coordinate	坐标	Derive	推导
Corner	角落	Descartes' rule of signs	笛卡尔符号规则
Correct; Right; Valid	正确	Descending	递减
Corresponding	对应的	Descriptive statistics	描述统计
Corresponding angles	同位角	Determinant	行列式
Cosecant	余割函数	Diagonal	对角
Cosine	余弦函数	Diameter	直径
Cotangent	余切函数	Dice	骰子
Count	数数	Difference	差；差别
Counter	相反的	Digit	数字；数位
Counter Clockwise	反时针	Digital	数码的
Cube	正方体	Dime	一毛钱
Cubic	立方	Dimension	维
Cubic root	立方根	Discriminant	判别式
Currency	货币	Disjoint	不相容的
Curve	曲线	Distance	距离
Customary	习惯	Distributive property	分配律
Cylinder	圆柱	Divergent	发散
		Divide	除
		Dividend	被除数
D		Divisibility	可除性
Data	数据	Divisible	可除的
Database	数据库	Division	除法
Dataset	数据集	Divisor	约数
Day	天	Divisor	除数
Day	日	Dodecahedron	十二面体
Decagon	十边形		

Dollar ($)	元；美元	Exchange, swap	互换
Domain	定义域	Exclusive	排斥的，独有的
Dot product	点积		
Dot; point	点	Expand	展开
Double	加倍	Expect	期望
Double	两倍	Experiment; Trial	实验
Double check	再检查	Exploration	探索
		Explore	探求
		Exponential, index	指数
E		Expression	表达式
Edge	棱	Extend	扩展
Eight	八	Exterior	外部
Eighteen	十八	Exterior angles	外角
Eighth	八分之一	Extra	多余
Eighth	第八		
Eighty	八十		
Einstein	爱因斯坦	**F**	
Elapsed time	过去的	Face	面
Element	元；元素	Fact	事实
Elementary	初级的	Factor ; divisor	因子
Elementary	基本的	Factorial	阶乘
Elementary School	小学	Factorial	阶乘
Eleven	十一	Factoring	因子分解
Elimination	消元 ；消除	Fahrenheit	华氏度
Ellipse	椭圆	Fallacy	悖谬
Ellipsoid	椭球	False	错
Endpoint	终点	Feet	英尺
Equal (=)	等于	Fewer than; Less than	比……少
Equation	方程		
Equidistant	等距	Fibonacci numbers	斐波纳契数列
Equilateral	等边多边形	Fifteen	十五
Equivalent	等效；等价	Fifth	五分之一
Error	误差	Fifth	第五
Estimate	估量；估计值	Fifty	五十
Estimation	估计	Figure; graph	图
Euler	欧拉	Find	找到
Evaluate	计值	Finger	手指
Evaluate	评估	Finite	有限
Even number	偶数	First	第一
Event	事件	Five	五
Exam	考试	Flip	翻转
Examine	检查	Foot ft)	英尺
Example；Instance	例子	For example; e.g.	例如

Formal	正式	Group	组
Formula	公式	Guess	猜
Formulate	规划		
Formulate; express	表达		
Forty	四十	**H**	
Four	四	Half	半
Four-vector	四维矢量	Half life	半衰期
Fourteen	十四	Halve	二分之一
Fourth	第四	Halving	对半分
Fourth; Quarter	四分之一	Harmonic sequence	调和数列
Fraction	分数	Height, tall	高
Fraction bar	分数线	Heptagon	七边形
Free fall	自由落体	Hexagon	六边形
Frequency	频率	Higher	更高
Frequency table	频率表	Histogram	直方图
Function	函数	Horizontal	水平的
Fundamental theorem of arithmetic	算术基本定理	Hour	小时
		Hour hand	时针
		Hundred	百
		Hundredth	第一百
		Hundredth; Percent	百分之一
G		Hypotenuse	斜边
Gallon (gal)	加仑	Hypothesis	假说
Gaussian distribution	高斯分布	Hypothesis testing	假设检验
Generate	生成		
Geometric	几何的		
Geometric sequence	等比数列	**I**	
Geometry	几何	Identical	相同的
Gram (g)	克	Identify	找出
Graph	画图	Identity	恒等式
Graph paper	坐标纸	Identity	恒等
Graphical	用图表示的	Imaginary number	虚数
Gravity	重力	Improper	不正当的
Greater	更大	Improper fraction	假分数
Greatest	最大的	Inch (in)	英寸
Greatest common divisor (GCD)	最大公约数	Incorrect	不正确
		Increase	增加
Greatest common factor (GCF)	最大公因子	Increasing	从小到大顺序
		Independent events	独立事件
Grid	方格	Independent variable	自变量
Group	分组	Indeterminate	未知数

Index	指数
India, Hindu	印度
Inequality	不等式
Infer	推断
Infinite	无限
Information	信息
Input	输入
Inside	里面
Integer	整数
Interior	内部
Interior angles	内角
Interpret; Explain; Justify	解释
Intersect	相交
Intersecting lines	相交线
Intersection set	交集
Invalid	无效
Inverse	反向
Inverse function	逆函数
Inverse operation	逆运算
Inverse property	逆向特性
Invert	倒转
Investigate; survey	调查
Irrational conjugate theorem	无理共轭定理
Irrational number	无理数
Irregular	不规则
Irrelevant	不相关
Isosceles triangle	等腰三角形

J

Joint probability	联合概率
Justify; Proof; Prove	证明

K

Key	关键
Key	重点
Key to a graph	图例
Kilogram (kg)	公斤
Kilometer (Km)	公里

L

Label	标记
Last	最后
Least common denominator (LCD)	最小公分母
Least common multiple (LCM)	最小公倍数
Least; Min; Minimum	最小
Length	长度
Less	更少
Less than (<)	少于
Like Denominators	公分母
Likely	可能的
Line	直线
Line graph	线型图
Line plot	折线图
Linear system	线性系统
List	列出
List	清单
Liter (L)	公升
Liter (L)	升
log	对数
Log-scale	对数尺度
Logarithm	对数
Logic	逻辑
Long division	长除法
Lorentz transformation	洛伦兹变换

M

Manipulation	处理
Manipulation	操作
Map	地图
Map; Corresponding	对应
Marginal probability	边际概率
Marks	标记
Mass	质量
Match	配对
Mathematics	数学

Matrix	矩阵	Multiplier	乘数
Max; Maximum	最大	Multiply; time	乘
Measure	测量		
Measure	量度		
Measure	度量	**N**	
Measurement	测量	Narrative	叙述
Median	中位数	Natural base	自然基数
Meter (m)	公尺	Natural logarithm	自然对数
Meter (m)	米	Negative	阴性
Method; Approach	方法	Negative number	负数
Metric system	公制	Negative; Minus	负
Metric units	公制单位	Neutron star	中子星
Micro	微	Next	下一个
Middle	中间	Nickel	五分钱
Midpoint	中点	Nine	九
Mile	英里	Nineteen	十九
Milliliter (ml)	毫升	Ninety	九十
Millimeter (mm)	毫米	Ninth	九分之一
Million	百万	Ninth	第九
Minkowski space	闵可夫斯基空间	Nonagon	九边形
		Nonstandard	非标准
Minuend	被减数	Normal distribution	正态分布
Minus	减；负	Not; No	不
Minute	分钟	Null	零
Mixed number	带分数	Null hypothesis	零假设
Mode	众数	Null; None	无
Model	模型	Number	数字
Modulo	取模	Number theory	数论
Mole	摩尔	Numeral	数位；数字系统
Momentum	动量		
Monomial	单项式	Numeration	计数方法
Month	月	Numerator	分子
Moore	摩尔	Numeric	数；数字的
More	更	Numeric expression	数字表达式
More than (>)	大于(>)		
Motion	运动		
Multiple	多个	**O**	
Multiple	倍数	Object	物体
Multiple Choice	多种选择	Obtuse angle	钝角
Multiplicand	被乘数	Obtuse triangle	钝角三角形
Multiplication	乘法	Octagon	八边形
Multiplicative inverse	倒数	Odd number	奇数
		One	一

Open	开放	Pint (pt)	品脱（pt）
Open circle	空心圆	Place value	位值
Operation	运算	Plane	平面
Order	次序	Plot	绘图
Order of operations	运算次序	Polar coordinates	极坐标
Ordinal number	序数	Polygon	多边形
Organize	组织	Polyhedron	多面体
Organized	有序	Polynomial	多项式
Origin	原点	Population	总体
Ounce (oz)	盎司	Positive	正；正确
Outcome	结果	Positive	阳性
Outlier	异常值	Positive number	正数
Output	输出	Possible	可能
Outside	外面	Post meridian (p.m.)	下午
		Pound (lb)	磅（lb）
		Power	幂
P		Power; Exponent	乘方
		Predict, prediction	预测
Parabola	抛物线	Preference	偏好
Paradox	悖论	Prefixes	前缀
Parallel	平行	Previous	前一个
Parallelogram	平行四边形	Primary school	小学
Parameter	参数	Prime factorization	质因子分解
Parity	奇偶性	Prime number	质数
Part	部分	Prism	棱镜
Pascal triangle	帕斯卡三角形	Probability	几率
Pass-code	密码	Probability	可能性
Pattern	模式	Probability	概率
Penny	一分钱	Probability	机率
Pentagon	五边形	Probability distribution	概率分布
Per	每个		
Percent	百分比	Problem	问题
Percentile	百分位	Process	过程
Perimeter	周长	Product	乘积
Period	周期	Product	积
Periodicity	周期性	Program; Application	程序
Permutation	排列		
Perpendicular	垂直	Programming	编程
pH	酸碱度	Proof by contradiction	反证法
Pi	派		
Pi	圆周率	Proper	正当的
Pictograph	象形图	Proper fraction	真分数
Pie chart	饼图	Property	特性

Proportional	正比
Protractor	分度规
Protractor	量角器
Pyramid	棱椎体
Pythagorean Theorem	勾股定理

Q

Quadrangle, Quadrilateral	四边形
Quadrant	象限
Quadrant	象限
Quadratic equation	二次方程
Quart	夸脱
Quarter hour	一刻钟
Quarter year; season	一季度
Quartile	四分位
Question	问；问题
Quiz	测试
Quotient	商

R

Radian (Rad)	弧度
Radical	根式
Radius	半径
Random	随机
Random	偶然
Range	范围
Range	值域
Range	值域
Range	极差
Rate	速率
Rate	率
Ratio	比；比率
Rational number	有理数
Rational root theorem	有理根定理
Rationale	依据
Ray	射线
Real number	实数
Reasonable	合理

Reciprocal	倒数
Rectangle	长方形
Rectangle	矩形
Rectangular Prism	长方体
Recurring	反复出现
Recurring	循环
Reduction	约分
Reduction; reduce	减少
Reference	参考；参照
Reference frame	坐标系
Reference frame	参考系
Reflect	反射
Reflex angle	优角
Regroup	重组
Regular, normal	正；正常的
Reject	拒绝
Relation	关系
Relativity	相对论
Relevant	相关
Remainder	余数
Repeat	重复
Repeating decimal	循环小数
Repetition	重复
Representation	表示
Result	结果
Review	回顾
Review	复习
Rhombus	斜方块
Rhombus	菱形
Right	右
Right angle	直角
Roman	罗马
Root	根，解
Rotation	旋转
Round down	下舍入
Round down	四舍
Round up	上舍入
Round up	五入
Rounding	四舍五入
Rule	规则
Ruler	尺

S

Sample	样本	Sixth	六分之一
Sample space	样本空间	Sixth	第六
Sampling	采样	Sixty	六十
Scalar	标量	Slope	斜坡；斜率
Scale	比例	Small	小
Scale	规模	Software	软件
Scalene triangle	不等边三角形	Solid	实
Scatter plot	散点图	Solution	解
Scientific notation	科学记数法	Solve	解；解答
Seasons	季节	Sort；Classification	分类
Secant	正割函数	Space	空间
Second	秒	Space-time	时空
Sector	扇形	Special case	特例
Segment	段	Specific	特定
Semester	学期	Specificity	特异性
Semi-straight line	半直线	Speed	速度
Semicircle	半圆	Speed	速度
Separately	分别	Sphere	球体
Separator	分隔号	Square	平方
Sequence	序列	Square	正方形
Sequence	数列	Square root	平方根
Series	系列	Standard	标准
Series	级数	Standard Deviation	标准差
Set	集合	Standard normal distribution	标准正态分布
Seven	七	Statement	陈述
Seventeen	十七	Statement	命题
Seventh	七分之一	Statistical power	统计功效
Seventh	第七	Statistics	统计
Seventy	七十	Step	阶梯，步骤
Shaded	阴影	Straight	直
Shape	形状	Straight angle	平角
Share	份	Subscript	下标
Show	说明	Subscript	下角义字
Side；Edge	边	Substitute	代入
Significant	显著	Substitute	替换
Significant figures	有效数字	Subtotal	部分和
Similar; Like	相似	Subtract; Minus	减
Simplify; reduce	简化	Subtraction	减法
Simultaneously	同时	Subtrahend	减数
Sine	正弦函数	Sum	和
Six	六	Summarize	总结
Sixteen	十六	Summary	概述

Summary	总结	Triangle	三角形
Superscript	上标	Triangular Prism	三角棱角
Superscript	上角文字	Trigonometric function	三角函数
Supplementary angles	补角	Trillion	万亿
Suppose	假如	Trillion	兆
Survey	测量	Trinomial	三项式
Symbols	符号	Triple	三倍
Symmetry	对称	True	是
		Turn	转
		Turn	圈

T

Table	表	Twelve	十二
Tail	尾	Twenty	二十
Tally	计数	Two	二
Tangent	正切函数	Type	种类
Technique	技术；技巧	Type I	第一类
Ten	十	Type II	第二类
Tenth	十分之一	Union set	并集
Tenth	第十		
Term	项		
Terminate	终结	**U**	
Theorem	定理	Unique	唯一
Third	三分之一	Unit	单位
Third	第三	Univariate	单变量
Thirteen	十三	Unlike denominators	异分母
Thirty	三十		
Thousand	千		
Thousandth	千分之一	**V**	
Thousandth	第一千	Valid	有效
Three	三	Value	数值
Three-dimensional	立体	Variable	变量
Time	时间	Variance	方差
Ton	吨	Vector	矢量
Top	上方；顶部	Vector	向量
Total	总和	Velocity	速度
Transformation	变换	Venn diagram	文氏图
Transistor	晶体管	Venn diagram	维恩图
Translate	平移	Venn diagram	温氏图
Transpose	转置	Verify	验证
Trapezoid	梯形	Vertical	垂直的
Tri-	三个	Vertical angles	对顶角
Trial and error	试错	Volume	体积

Volume   容量

W
Week   星期
Whole   整体
Whole number   整数
Width   宽；宽度

Y
Yang Hui triangle   杨辉三角
Yard   码

Z
Zero   零；零点

索引 | Alphabetical Index

www.ingramcontent.com/pod-product-compliance
Lightning Source LLC
Chambersburg PA
CBHW072025060426
42449CB00035B/2598

9 781963 384017